PRACTICAL GUIDE TO QUALITY POWER FOR SENSITIVE ELECTRONIC EQUIPMENT,
2nd EDITION

By the Editors of *EC&M*

Edited by R. M. (Ray) Waggoner
Consulting Editor *EC&M*

Intertec Publishing Corp.
Overland Park, Kansas

Practical Guide To Quality Power For Sensitive Electronic Equipment, 2nd Edition
Copyright © 1997 Intertec Publishing Corporation
All rights reserved

First Printing: August 1997

Published by
EC&M Books
Intertec Publishing Corporation
9800 Metcalf Avenue
Overland Park, KS 66212-2215

Library of Congress Catalog Card Number: 97-74773
ISBN 0-87288-667-0

Please Note: The designations "National Electrical Code," "NE Code," and "NEC," where used in this book, refer to the National Electrical Code® which is a registered trademark of the National Fire Protection Association.

CONTENTS

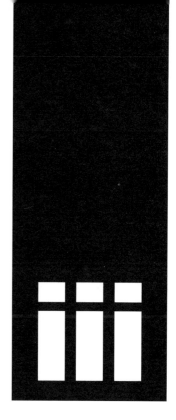

Preface vii

CHAPTER 1 — Quality Power Basics 1
Reason For Increased Sensitivity 1
FIPS Publication 94 2
The CBEMA Curve 2
Using The Curve 2
Typical Analysis 3
New ITIC Curve 4
IEE Emerald Book 5
Sine Wave Distortion 7
Other Wave Shapes 7
Nonlinear Loads 8
Power Disturbances 8
DC Power Supplies 8
Impact of Voltage Sag 9
Impact of Overvoltage, Surges, RFI, and Harmonics 9

CHAPTER 2 — Nonlinear Loads and Harmonics 11
The Effect of Loads 11
Distortion of Current Sine Waves 11
Harmonics 12
Voltage Wave Distortion 13
Effects of Harmonics on Electrical Loads 13
Overheated Neutrals 15
Harmonics and Power Factor 16
Harmonic Resonance 17
Major Source of Harmonics 17

CHAPTER 3 — Defeating Harmonics 19
General Guidelines 19
k-Factor Transformers for Harmonic Loads 20
Derating Existing Transformers 21
Zig-Zag Transformers 22

Sizing Generator Sets for Nonlinear Loads 22
Coping with Generator Harmonics 23
Variable-Speed Drives and Harmonics 24
VSD Problems and Solutions 25
VSD Solutions 26
Harmonic Filters 26
Other Solutions 28

CHAPTER 4 — Other Factors Affecting Solid-State Electronic Devices 31
Effects of Noise 31
Types of Noise 31
Common-Mode Noise 32
Ground Loops 33
Normal-Mode Noise 33
Types of Transients 33
Transients Caused by Lightning and Switching 35
RFI/EMI Noise 36
Power Disturbances 36
Long-Time Voltage Aberrations 37
Intermediate Voltage Aberrations 39
Short-Time Voltage Aberrations 39
Frequency Aberrations 39
Input Power Guidelines 39

CHAPTER 5 — Defeating Noise and Other Problems 41
Coping with Noise 41
Other Noise-Reduction Approaches 42
Defeating Noise on Wire Pathways 42
Defeating Induced Noise 43
Defeating Radiative Noise 44
Avoiding Ground Loops 45
IG-Type Wiring 45
IG Wiring Methods 46
Other IG Wiring Tips 47
Power Conditioning 48
Increased Service Reliability 49
Superconductivity 51

CHAPTER 6 — Grounding for Sensitive (Solid-State) Equipment 53
Importance of Grounding 53
Effective Computer Grounding 54
Power System Grounding 54
High-Frequency Grounding 55
Establishing an SRG 56
Other Items Grounded to SRG 57
Avoiding Grounding System Noise 59
Grounding Electrodes 59
System Resistance 60
Ground Rods 60
Reducing Resistance 61

CHAPTER 7 — Surge and Transient Protection 63
Lightning 63
Surges Induced by Equipment 65
Surge Arresters 65
Locating Surge Arresters 65
Sizing Lightning Surge Arresters 66

Lightning Surge Arrester Lead Length 67
Protecting Solid-State Devices 67
Transient Suppressors 68
Attenuation Devices 70
Applying TVSS Protection 70
Selecting TVSS Equipment 71

CHAPTER 8 — Power Conditioning and Reliability 73

Voltage Regulators and Line Conditioner 73
Constant-Voltage Transformers 74
Tap-Switching Regulators 74
Line Conditioners 75
Shielded Isolation Transformers 75
MG sets 75
Magnetic Synthesizers 79
Static UPS Systems 79
Static UPS Configurations 80
Other Static UPS Considerations 81
Static UPS Redundancy 82
Rotary UPS Systems 83
Hybrid Rotary UPS 84

CHAPTER 9— Site Analysis and Monitoring 85

Determining Power Quality 85
Inital Steps 85
Inital Testing for Harmonics 86
Advanced Tests 87
Waveform as a Diagnostic Tool 88
Computer Analysis 88
CBEMA (ITIC) Curve Analysis 89
Measurements 90
Practical Tips 90
Identifying Grounding Mystery Disturbances 90
Locating Utility Mystery Problems 91
Locating Harmonic Mystery Problems 92

CHAPTER 10— Troubleshooting Techniques 95

Noise in Data Communications Cable 95
Diagnostic Instruments 96
Communication Cable Noise Example 98
Decibels and Electrical Measurements 100
Sample dB Calculation 101
CCTV Signal Attenuation Example 101
Isolation Transformer Attenuation Example 103
Example, Locating an EMI Source 104

CHAPTER 11— Design and Installation Considerations 107

Interaction and Varying Impedances 107
Effect of Impedance on THD 109
Capacitance and Resulting Resonance 109
Harmonics and Power Factor 111
Triplen Harmonics and Neutral Overloading 112
Cable Derating and Nonlinear Load Panelboards 112
Computer Signal-Reference Ground 113
Grounding Instrumentation and Controls 115
Hints on Installing Electronic Cash Registers 116

CHAPTER 12— Equipment Case Histories 119
Harmonic Filters 119
MG Sets 119
Subcycle Static Transfer Switches 121
Surge Protection 122
Power Distribution Units 124
Disturbance Analyzers 125

CHAPTER 13— Location Case Histories 129
Overnight Shipper 129
Frozen Food Processor 130
Telecommunications Manufacturer 131
Office Building 131
Die Maker 132
Test Facility 133
Computer Company 134
Automated Warehouse 134
Hospital 135
Mechanical Contractor 136
Medical Supply Company 136

CHAPTER 14— Coordination Guidelines for Sensitive Electronic Loads 139
Continuity of Processing Operations 139
Facility Location Considerations 139
Commercial Power Considerations 140
Coordination and Planning with CPU Vendor 140
Designing the System 141
Matching System Power Requirements to Power Conditioning Alternatives 142
Grounding for Consistent Noise Suppression 142
Redundancy Requirements 143
Data Communications Protection 143
Lightning Protection 143

PREFACE

This second edition of the "Practical Guide to Quality Power" is a thoroughly updated version of the original. Lots of new information, techniques, and equipment have become available since the first edition was published in 1993. Rather than only inserting new material into the existing chapters, the book has been completely overhauled and greatly expanded. Several new chapters have been added in order to include case histories that illustrate some of the concepts covered in the book.

Sensitive electronic equipment is found in almost all business offices, factories, commercial establishments, health care facilities, school campuses, and endless other locations. This category of equipment includes not only computers and peripherals used for data-processing, but also mainframe computers serving interconnected distributed computers; telecommunication equipment; process control computers; and electronic control and instrumentation. So central to economic wellbeing have these items become, that a great deal of effort focuses on keeping them up and running efficiently.

The most important need of sensitive equipment is a dependable source of electric energy, a grounding system that provides a stable platform for consistant operation, and protection against transient surges. These and other factors must be addressed at several levels if the effort is to be successful.

First, the location in which the electronic devices are installed must be planned carefully to assure *essential infrastructure* is in place. This phase of the requirement is often overlooked or poorly understood. A poorly designed electrical power distribution system will act as a highway for transients to the sensitive equipment, and allow powerline surges and outages to affect operations. A companion book to this one entitled, *Practical Guide to Power Distribution Systems For Information Technology Equipment* covers what you must know about the subject.

This book, on the other hand, covers the second phase of filling the needs of sensitive or solid-state electronic equipment. The emphasis here is on the *quality of power* that is fed to this equipment.

What "quality of power" refers to is whether: the voltage and current are pure sine waves, or contain harmonics that distort the waves; voltage and current are within the upper and lower limits; transients have been eliminated; the grounding of equipment provides protection for personnel against injury and allows signals to be transmitted accurately; and other similar considerations.

The order in which the various subtopics of quality power are presented does not necessarily follow the way the subject would be taught in a classroom. For instance, nonlinear loads and harmonics is covered in Chapter 2, while from the teaching standpoint it might be more logical to first discuss voltage stability, grounding, and similar items that affect solid-state devices.

The reason for this is the reader should first read and understand the material covered in the companion book to this one. Another reason is that nonlinear loads and harmonics are the most pressing problems facing persons designing an installation including electronic equipment, or operating and maintaining the devices.

This book provides a discussion of the various power quality problems affecting solid-state devices and ways in which these can be defeated, either by using better wiring and grounding techniques or by applying power conditioning and/or surge protective equipment.

Alfred Berutti, PE
Overland Park, KS
Editorial Projects Consultant
August 1997

QUALITY POWER BASICS

When the subject of power quality is discussed, the mistaken assumption is often made that the topic only has to do with computers. At one time this may have been true since data-processing computers were among the first significant loads that did not always operate reliably on the raw power received from the serving electrical utility. With the advent of the microprocessor, however, there is a host of equipment that operates at voltage levels similar to that of the data processing computer. Items as diverse as electronic instrumentation, cash registers, scanners, motor drives, telecommunication equipment, and others, all depend upon onboard "brains" to give them instructions. Thus, the quality of power this equipment receives is as important as that supplied to computers. The broad category that covers all such equipment, including personal computers, mainframe, and process-control computers, is referred to in this book as both "solid-state electronic devices" and "sensitive electronic equipment."

REASON FOR INCREASED SENSITIVITY

The heart of the problem that seems to have suddenly appeared is that while the upper limit of circuit speed of modern digital devices is continuously being raised, the logic voltages have simultaneously been reduced. Such a relationship is not accidental. As more transistors and other devices are packed together onto the same surface area, the spacing between them is lessened. This reduced distance between components helps lower the time the circuit requires to operate. On the other hand, if the circuit voltage remained the same, the closer packing would increase the probability the current would seek a shorter path to ground due to insulation failure between two adjacent components. The way of preventing this is by reducing the circuit voltage.

Many advantages have resulted from reduced power and logic-circuit volt-

Sensitive (solid-state) electronic equipment *includes not only data-processing computers, but also instrumentation and other microprocessor-dependent items.*

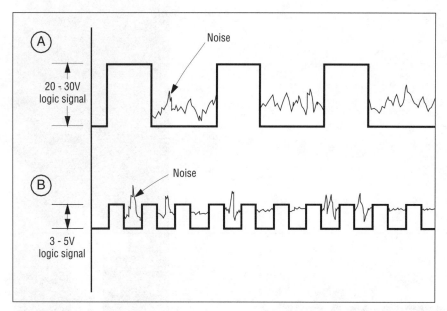

Fig. 1.1. Noise riding within a square wave (**A**) has no significant effect as long as the signal voltage is high. With reduced signal strength (**B**), the noise may now cause false triggering of sensitive electronic loads.

age. Where the original computers consisted of rooms full of vacuum tubes that gave off a large amount of heat, a few microprocessors with virtually the same capacity have vastly lower power consumption and cooling requirements. The down side is that these circuits are much more susceptible to the same fast noise pulses that have always existed on utility power lines.

Fig. 1.1 shows the reason for the increased sensitivity. With the higher signal voltage levels in (**A**), the noise pulses are swamped by the logic signal voltage. With this high signal-to-noise ratio, the circuit responds only to the appropriate signal.

In (**B**), the ON and OFF (0 and 1) digital logic signals have been speeded up and at the same time the voltage level has been lowered. With the signal strength reduced from the 20-30V level to 3-5V, the same level of noise as previously shown now becomes a factor. The logic circuit can interpret noise-voltage spikes as being legitimate ON/OFF signals, causing erroneous data to be processed, or refusing to accept the signal since it does not meet a "parity" check.

Actually, the situation is even more critical. Logic circuits have now been operated at between 1/2-1V. Even what used to be considered to be minor noise is now a potential threat to the correct operation of the system. The challenge then is to reduce the noise conditions or to manipulate them in such a way that they will not interfere with the operation of solid-state electronics devices.

FIPS PUBLICATION 94

For persons who need more information on the general subject of the quality power requirements of solid-state devices, the best point to begin is the FIPS Publication 94. In 1983, the U.S. Department of Commerce published a guide summarizing the fundamentals of powering, grounding, and protecting sensitive (solid-state) electronic devices. The document is known as FIPS Pub 94, *Federal Information Processing Standards Publication*. While some of the information has been outdated by the rapid changes in the industry, the majority is still valid and is explained in as easy-to-understand language as possible.

After being written, the guideline was sent to the Computer Business Equipment Manufacturers Association (CBEMA) to find out if the language used in the publication was appropriate. When the CBEMA group put its stamp of approval on the information, the whole industry finally had an overarching guideline.

As originally conceived, FIPS Pub 94 was written to cover automatic data processing equipment (ADP), which at that time constituted the principal equipment experiencing difficulty running on normal utility-supplied power. This, however, does not mean that FIPS Pub 94 has been obsoleted. It just needs to be read with the understanding that any reference to ADP in the publication should be taken to mean any device or equipment depending upon microprocessors for its operation.

FIPS Pub 94 is a guideline intended to provide a cost/benefit course of action. As a result, it can be relied upon to give the best solution to problems that will be encountered, for the least amount of money. It serves as well as an indication of how we got into the "sensitivity crisis" and in providing the fundamentals of how to get out of the dilemma.

THE CBEMA CURVE

In addition to approving the FIPS No. 94 document, the CBEMA group provided a curve to be included in the publication that had been used as a guideline for their members in designing their power supplies. This curve was designed by several in the computer industry to point out ways in which system reliability could be provided for their electronic equipment. This original "CBEMA Curve" from the FIPS document is shown in **Fig. 1.2**.

The curve is a susceptibility profile. In order to better explain its meaning here, the curve has been simplified and redrawn as **Fig. 1.3**. The vertical axis of the graph is the percent of voltage that is applied to the power circuit, and the horizontal axis is the time factor involved (in microseconds to seconds). In the center of the chart is an acceptable area, and on the outside of that area is a danger area on top and bottom. The danger zone at the top involves tolerance of equipment to excessive voltage levels. The danger zone on the bottom sets the tolerance of the equipment to a loss or reduction in applied power. This guideline says that if the voltage supply stays within the acceptable area, the solid-state equipment will operate well.

USING THE CURVE

Computers, programmable logic controllers, distributed processing units, instrumentation, telecom, and other

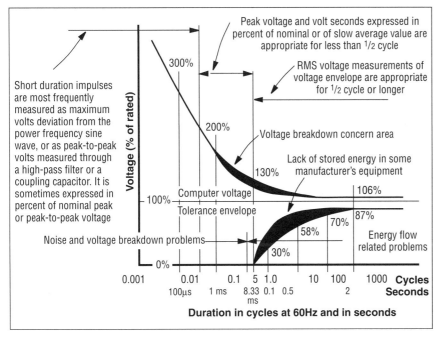

Fig. 1.2. The CBEMA curve from FIPS Pub 94.

solid-state systems will operate reliably if applied carefully. All these units have one thing in common—they are voltage and time sensitive. Voltage sags and swells, as well as outages and transients will seriously affect their operation.

Many problems that affect solid-state electronic equipment are first noted as mystery disturbances. For example, a piece of solid-state equipment in some cases seems to react to a blink of the lights. Sometimes when this occurs, there is absolutely no noticeable disturbance to the power distribution system. These kinds of disturbances are the most troublesome because they occur in a very fast time frame on the power wiring, or may have occurred on the ground and neutral wiring in a facility and be unrelated to any power source disturbances.

Another group of disturbances that might be encountered is caused by the supply voltage being too high or too low for the operating characteristics of the solid-state equipment. Also, voltage variations or transients caused by other equipment within the facility, or even from disturbances on neighboring power systems, could be the source of the problem.

In addition, a solid-state electronic load will shut itself down after a certain period of power instability to protect its hardware and logic system.

These types of disturbances can lead to the conclusion that some form of uninterruptible power source, perhaps with standby generation, is required. Before embarking on such a course, however, it is best to analyze the pattern of the disturbances against the CBEMA curve to see what the solid-state equipment can tolerate. Based upon the results, a solution that will most effectively deal with the problem can be worked out.

In order to use the curve, it is first necessary to determine what are the natures of the disturbances that are most prevalent in the facility.

The first step is to use specialized instruments. Disturbances associated with powering, grounding, and protecting solid-state devices can be measured, analyzed and evaluated using test instruments specifically intended for digital logic systems. These instruments, when located near the suspected disturbance, or when measuring the unusual operation of the power distribution system, will provide data on voltage fluctuations, short- and long-time excursions, and the specifics of how the disturbance places the equipment at risk.

Once these measurements have been taken, preferably with recording-type instruments, the results can be analyzed in combination with the CBEMA curve to help understand the nature of the disturbances.

TYPICAL ANALYSIS

There are various types of power disturbances that can affect electronic equipment. The CBEMA curve helps understand how each type that has been recorded will influence the operation of the system.

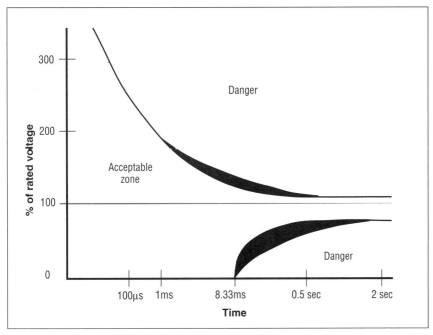

Fig. 1.3. A simplified version of the CBEMA curve. Voltage levels outside the acceptable zone results in system shutdown and potential hardware and software loss.

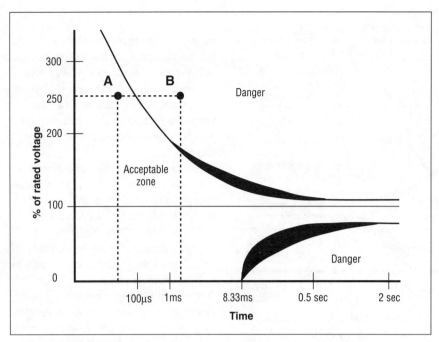

Fig. 1.4. Short-duration transients are acceptable to sensitive electronic equipment as denoted by Point A; longer ones cause disruptions as denoted by Point B.

Transients. In **Fig. 1.4**, a short term disruption known as a transient (dot A) reaches a maximum value of 250% of normal voltage and is very fast (less than 1 microsecond). Thus, it falls within the curve and is tolerable. But when the same level of transient (dot B) occurs for a longer period of time, it will cause a disruption.

Short outages. The effect of disturbances that last a longer time than transients can also be determined by applying the CBEMA curve (see **Fig. 1.5**). There are two interesting areas to look at: they are to the left and right of the 8.3 millisecond mark on the bottom of the curve. To the left is a zero-voltage area where this outage is acceptable; to the right is the danger zone where voltage at a certain minimum voltage must have been restored to the device or a shutdown will occur. The voltage needs to be pretty close to the normal operating tolerances by 1/2 sec (30 electrical cycles). Notice that to the left hand side of the 8.3 ms mark the equipment can be run with no voltage!

That may sound facetious. Why talk about such a short time span—8 ms? A blink of the eye takes longer than that! That is correct, but 8.3 ms is a lifetime for much of the solid-state electronic equipment being used today. It is important to recognize that 8.3 ms is a time frame in which you can work. In the example, a power outage lasting 4 ms is acceptable, while one lasting 70 ms would cause a shutdown.

For example, a subcycle solid-state switch operates to transfer between two power sources in 4 ms or less. If that switch transferred back and forth from one source to another, the sensitive equipment would not be disturbed because that point is within the tolerance envelope. However, even the fastest utility or service-entrance circuit breaker available (70 ms) is likely to cause a shutdown once it opens.

Long-term disturbances. Most of the disturbances that are likely to be encountered (in the neighborhood of 95%) have cleared up by the 1/2-second mark. Thus, when a UPS system (capable of taking care of all interruptions for perhaps 15 to 30 minutes or longer) is used to solve a particular problem, it may be overkill. Providing such an expensive system to do the job of protecting against problems lasting 1/2 second or less may be considerably more than is required. Lower cost protection might be able to be utilized!

The CBEMA curve helps sort out the kind of difficulties that are occurring at the jobsite, where they fall within the curve as measured by power disturbance instruments, and then how to select the most effective and economical mitigation techniques to assure dependable operation of the electronic equipment.

NEW ITIC CURVE

The sponsoring organization of the CBEMA curve has changed its name to the "Information Technology Industry Council" (ITIC). A working group

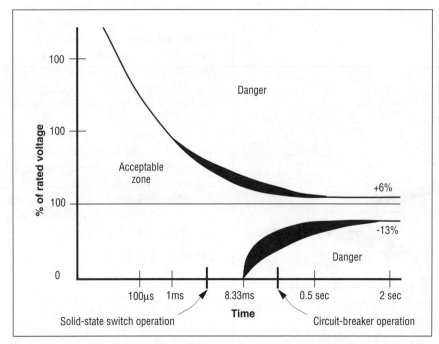

Fig. 1.5. A solid-state switch can operate within typical sensitive equipment tolerance, while a utility circuit breaker operates 10 times slower than is tolerable.

of that organization in collaboration with EPRI's Power Electronics Applications Center (PEAC) has taken on the task of developing a new version that more accurately reflects the performance of typical single-phase, 60 Hz computers and their peripherals, and other information technology items such as copiers, fax machines, and point-of-sale terminals. The new ITIC curve is shown in **Fig. 1.6**. While specifically applicable to computer-type equipment (as with the CBEMA curve), it is generally applicable to other equipment containing solid-state devices.

Because this new ITIC curve has some carefully negotiated data points, each of these points will be useful as criteria to test the performance of a given product. For the vertical percent-of-nominal-voltage point selected, if the output is not affected until later than the value shown on the horizontal time scale for that point, the product has met the limit described by the curve.

It should be understood, however, that this new curve is not intended to reflect the performance of all electronic-based equipment. There are too many variables such as power supply loading, nominal operating voltage level, and process complexity to try to apply a "one size fits all" ITIC curve. At present, there are plans to develop an international version of the curve that will reflect the performance of 230V, 50 Hz equipment. PEAC is simultaneously working to develop an ITIC-type curve for adjustable-speed drives.

IEEE EMERALD BOOK

A recent addition to the Institute of Electrical and Electronic Engineers (IEEE) color book series, IEEE Standard 1100 (Emerald Book), *Recommended Practice for Powering and Grounding Sensitive Electronic Equipment*, seeks to bring order to power quality assurance, particularly in industrial and commercial power systems. It covers much the same ground as FIPS No. 94, but in greater engineering detail. Included in the book is an updated

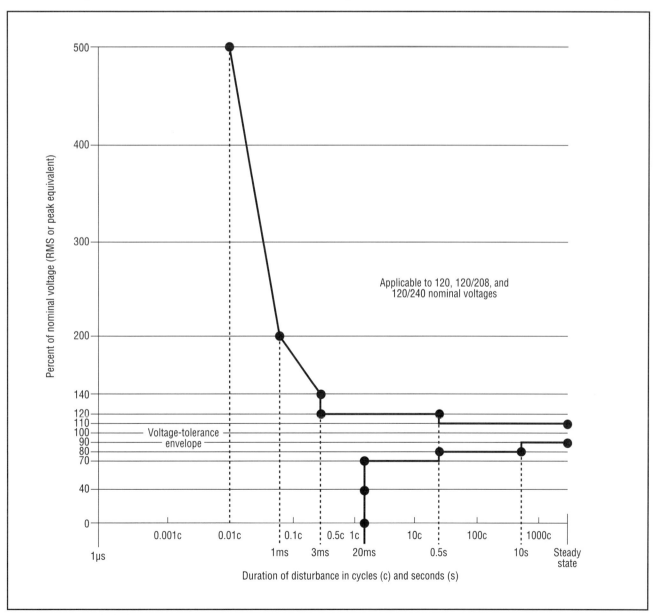

Fig. 1.6. *The new ITIC (CBEMA) curve.*

version of the CBEMA/ITIC curve.

The Emerald Book is divided into nine chapters.

Chapter 1 is the introduction; it describes the scope and background of the standard. The highlights of the other eight chapters are as follows.

Chapter 2 contains definitions. The power quality industry suffers from a lack of consistent terms, with definitions taking on many meanings. This lack of common terminology inhibits clear communication and promotes confusion. Here is a sampling of the IEEE definitions.

Power Quality. "The concept of powering and grounding sensitive electronic equipment in a manner that is suitable to the operation of that equipment."

Stated another way, a power quality problem results when there is a difference between the quality of power supplied and the quality of power required for reliable load equipment operation. When viewed in this context, power quality problems can be solved by: reducing the magnitude or frequency of occurrence of power supply variations, including wiring and grounding problems; improving the tolerance of the load equipment to the power supply variations; or adding power conditioning equipment between the power supply and sensitive load equipment to mitigate the power supply variations to a level acceptable to the load equipment.

Ground. "A conducting connection, whether intentional or accidental, by which an electric circuit or equipment is connected to the earth, or to some conducting body of relatively large extent that serves in place of the earth."

It's important to note that ground connections can be intentional or accidental. Further, "ground" is not necessarily the "earth." For example, the airframe becomes the ground for an airplane. Likewise, the surroundings (building steel) on the 40th floor of a high-rise becomes "ground" rather than the earth 40 floors below.

Ground Loop. "A potentially detrimental loop formed when two or more points in an electrical system that are nominally at ground potential are connected by a conducting path such that either or both points are not at the same ground potential."

Note that a ground loop is only *potentially* detrimental. In subsequent chapters of the Emerald Book, you'll find that ground loops may be detrimental, good, or benign. For example, when analog 4-20mA signals are involved, the ground loop currents will cause false readings. On the other hand, digital circuits where higher frequencies are of interest, multiple ground connections (with ground loops) become necessary to equalize potentials at high frequencies. In other instances, ground loops have no effect.

Isolated Equipment Ground. "An insulated equipment grounding conductor run in the same conduit or raceway as the supply conductors...originates at an isolated ground type receptacle or equipment input terminal block and terminates at the point where the neutral and ground are bonded at the power source."

Contrary to its name, an isolated ground (IG) is not isolated from the power system ground. Rather, it is insulated at the receptacle and intermediate panels to control where the connection to the power system ground is made. An IG is allowed by exception and should be used only where required to reduce electrical noise. In certain interconnected systems, IG has been found to increase the noise on the interconnecting signal cabling and should be avoided in most instances.

Note that where IG is used in this "Quality Power" book, it will always be referred to as "insulated ground" rather than "isolated ground" as defined in the Emerald Book. For further information on how IG circuits are to be applied and installed, see Chapter 5 in this book.

Power disturbancd. "Any deviation from the nominal value (or from some selected thresholds based on load tolerance) of the input AC power characteristics."

With powerline monitoring equipment, it's possible to measure and record very slight changes in any of a number of characteristics of the AC power supply. Just because a deviation from the ideal nominal value can be detected does not mean that it's a power disturbance; the load equipment may or may not be affected by the power deviation. It's proposed that the term "power disturbance" be used to indicate only those variations that actually disturb the load.

The Emerald Book also lists words that have a history of varied usage, some without specific meaning. These terms are recommended to be avoided and include "blackout," "clean ground," "computer grade ground," "dedicated ground," "dirty ground," "dirty power," "glitch," "raw power," and "spike."

Chapter 3 discusses the nature and origin of power quality problems. It describes the origin of power supply variations and the typical sensitivities of load equipment. Lightning is an obvious source of transient voltage surges and also a major source of voltage sags as the overvoltages cause flashovers and faults in the power system. Load switching is identified as another major cause of disturbances.

Chapter 4 reviews the fundamental concepts related to power quality to help explain the "why's" of the recommended practices presented in Chapter 9 of the Emerald Book. Topics include impedance considerations, frequency effects, load and power source interactions, harmonic current effects, sources and characteristics of voltage surges, coupling mechanisms, and grounding, bonding, and shielding fundamentals.

Grounding requirements for sensitive electronic equipment are divided into distinct but interconnected subsystems: safety (fault and personnel protection) subsystem following the NEC requirements; signal reference subsystem for equipment performance by maintaining equal ground potentials over a broad range of frequencies; and lightning protection subsystem.

Chapter 5 describes 18 types of instruments that are useful in performing power quality site surveys. Some of the important lessons of this chapter are: use the right instrument for the task; know the characteristics of the instrument; and always use true-rms instruments when measuring nonsinusoidal waveforms.

Chapter 6 is a primer on conducting site surveys, including forms and

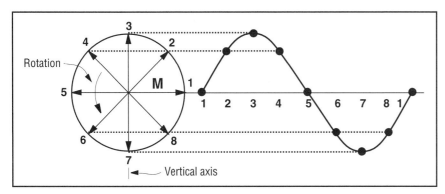

Fig. 1.7. Generating a sine wave.

procedures, what to look out for, and typical problems encountered. A thorough investigation of the site wiring and grounding is recommended when troubleshooting power quality problems before resorting to the more expensive and time-consuming task of power line monitoring. Careful testing, troubleshooting, and documentation techniques are necessary to collect meaningful power quality data.

Chapter 7 briefly describes 25 problems and solutions organized into nine typical problem categories, including utility problems, premises-generated surges, electronic load problems, premises wiring problems, TVSS problems, radiated electromagnetic interference (EMI) problems, electrical inspection problems, life-safety system problems, and equipment application problems.

Chapter 8 describes commonly used power conditioning equipment. The chapter also includes a discussion on writing and reading product specifications, with particular emphasis on uninterruptible power supply (UPS) product specifications.

Chapter 9 is the heart of the book, with the recommended power and grounding practices organized by power system components from the service entrance to the outlet box. The recommended practices include:

• strictly following the requirements of the NEC;

• using solidly grounded AC power systems; using dedicated circuits for sensitive loads;

• using an insulated grounding conductor to supplement the Code-minimum raceway grounding path; and

• using a separately derived source close to the sensitive loads.

Presently, an IEEE working group is revising the Emerald Book. Proposed revisions include updating all chapters, adding more figures for clarification, and adding a new chapter specifically addressing recommended practices for distributed computer and telecommunications equipment. Approval is scheduled for the last quarter of 1997, and issue in early 1998.

SINE WAVE DISTORTION

To adequately understand the subject of power quality, it is important to understand the nature of "clean" or "pure" power. A clean power system describes a system where the current and voltage waveforms are pure sinusoids. A sinusoid or sine wave is the plot over time of the trace a vector (M) makes when rotating from a 0° position.

The shape of a sine wave can be produced graphically as shown in **Fig. 1.7**. Draw a circle whose radius (M) is equal to the maximum value of the sine wave. Divide its circumference into a number of equal parts (in this example, 8). Through the center of the circle draw a horizontal line and divide it into the same number of equal parts. The horizontal divisions represent time. When the vector M is pointing towards 1, on the time scale, it is neither above nor below the horizontal line. When the vector M rotates counterclockwise to position 2, its height above the horizontal line is projected over to point 2 on the time scale. When at point 3, vector M has its maximum positive value. This is continued until vector M once again reaches point 1. When all the points on the time scale are connected by a smooth curve, the result is a sine wave. The more segments the circle is divided into, the more nearly it appears as a sine wave on the time scale.

The time required for the vector M to make one complete revolution (a cycle) is called its period. The example sinusoidal waveform, therefore, has a frequency equal to one period. Frequency is maintained at 60 cycles per second (60 Hz) in the US; in some other parts of the world, it is 50 Hz. Because this waveform repeats at regular intervals (every 360°), it is referred to as periodic.

OTHER WAVE SHAPES

Waveshapes having a frequency that is different than the fundamental 60 Hz sign wave of a power system also exist. When this waveshape is superimposed on the 60 Hz waveshape, it will add to the fundamental at each point along the waveshape and produce a distortion as shown in **Fig. 1.8**.

In this example, the second waveshape has a frequency that is three times that of the fundamental (180 Hz). This is called a "third harmonic." When the two waveshapes are added together at each point, the resultant is a characteristic "flat-top" waveshape that is the signature of a sine wave having third harmonic content with an in-phase relationship.

A fifth harmonic (300 Hz) repeats

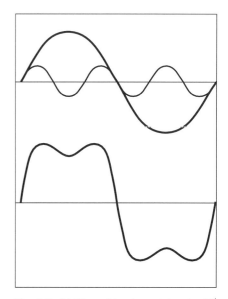

Fig. 1.8. Addition of fundamental and a 3rd harmonic produces a distinctive wave-form distortion.

five times every cycle. Adding it to the fundamental would distort the waveshape in its own unique way. So do all the other odd harmonics (7, 9, 11, etc.).

Waveshape distortion is made an even more complex subject by its variability. For example, the same harmonic that is not synchronized with the fundamental as shown previously, will give an entirely different resultant waveshape, as seen in **Fig. 1.9**. The picture is further complicated by the fact that more than one harmonic may be present at the same time (say the 5th and 7th). When they are added together with the fundamental, the resultant waveshape is once again different. The subject of harmonics and how they distort the sine wave is covered in Chapter 2 of this book.

No real current or voltage waveshape is a pure sine wave; it is always somewhat distorted. Certain deviations can be reproduced exactly by adding together a series of sine waves of a particular frequency, amplitude, and timing. Nonsinusoidal waveforms, therefore, can be broken down into some finite number of pure sine waves. This knowledge has been put to good use in analyzing a circuit by using diagnostic instruments. With these modern tools, it is possible to analyze a waveshape and identify what harmonics are present. A further explanation of the techniques used is given in Chapter 9.

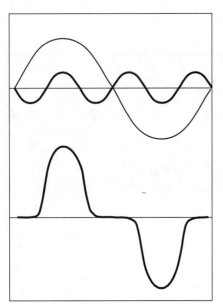

Fig. 1.9. If the same 3rd harmonic starts in the other direction, the resulting wave-form is totally different.

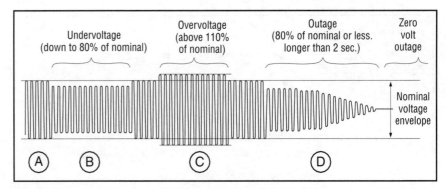

Fig. 1.10. Normal voltage (A), sag (B), swell (C), and outage (D).

NONLINEAR LOADS

The most common sine waves that distort a power system waveshape are whole number multiples of the fundamental 60-Hz power frequency. These multiples are called harmonics. For example, the second harmonic's frequency is 120 HZ, the third harmonic's is 180 Hz, and so on.

Traditional linear loads such as motors, heaters, capacitors, etc., allow voltages and currents of the fundamental frequency to appear in the power system with little or no harmonic distortion. However, nonlinear loads, such as switching power supplies, power thyristors (SCRs), electronic ballasts, and others, can introduce significant levels of harmonics into the system. The result will be a sine wave distorted by the addition of a harmonic as explained previously.

The negative effects of nonlinear loads are not felt directly by most sensitive electronic equipment because they are isolated from the power line by their DC power supplies. However, excessive distortion of the incoming power waveshapes can make it unacceptable to the power supply or other equipment and cause a shutdown.

Because of the importance of the nonlinear load problem, the subject is dealt with at length in Chapter 2.

POWER DISTURBANCES

Another major cause of erratic operation of solid-state electronic equipment is disturbances on the power line, such as transient spikes, voltage fluctuations, and outages.

There are several terms used to describe whether an electrical power system is operating normally or not. A waveform illustrating the different disturbances is shown in **Fig. 1.10**.

Long-term under/overvoltage can last for more than a few seconds and are usually caused by circuit overloads, poor voltage regulation, or utility brownouts. They adversely affect the operation of sensitive electronic equipment.

Sags are decreases in voltage (less than 2 second duration) outside the normal tolerance of the electronic equipment. They are often caused by the starting of heavy loads, or faults occurring in the power system. Sags represent momentary depletions in available energy that can cause equipment shutdown.

Swells are increases in voltage (less than 2 second duration) outside the normal tolerance of the electronic equipment. They are usually generated by sudden load decreases such as the deenergizing of a very large motor. They can damage electronic equipment having insufficient overvoltage tolerance.

Outage is the complete loss of power lasting from several milliseconds to several hours. They are usually caused by power system faults, accidents involving power lines, and equipment failure. Long-term outages affect all unprotected electrical and sensitive electronic equipment.

DC POWER SUPPLIES

Virtually all modern solid-state equipment runs on the DC power it receives from a complex electronic power supply. Many electronic equipment manufacturers assume their hardware will operate from a distribution network with zero internal impedance, receive

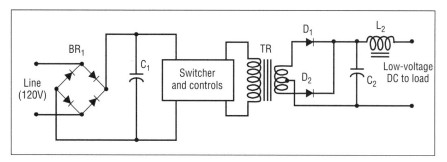

Fig. 1.11. Simplified block diagram of a switching-mode power supply.

a pure sine-wave, and never be subject to line voltage variances of ±5% from nominal. As such, most power supplies are designed to accommodate prolonged line voltage sags to these levels. However, the combination of utility and locally generated disturbances knows no such modest bounds. Many of the problems in operating solid-state electronic equipment are due to the incompatibility of the power supply with the voltage it is receiving.

As shown in **Fig. 1.11**, the DC power supply is a sophisticated piece of equipment. Its basic function is to deliver stabilized low-voltage DC to the digital logic it feeds. Based on a fast-switching DC-to-DC converter, the unit converts rectified 60 Hz into the low-voltage DC (typically 5V) required by computer logic. The nominal 5VDC output is compared to an accurate 5V reference so that an error-correcting feedback signal can be developed. This signal adjusts the relative ON and OFF durations of the DC-to-DC converter, thereby holding the output at the required 5V.

In principle, switch-mode power supplies (SMPSs) have the ability to bridge a total power outage for periods up to three complete cycles. The key requirement for maximum immunity against power outage lies in the filter capacitor (labeled "C_1" in the diagram) being fully charged to design voltage. This capacitor acts like a short-term battery; during a power outage, the power supply's DC-to-DC converter is kept running by drawing current from this capacitor.

Stored capacitor energy is equal to one half of capacitance times peak voltage squared [$½C \times (V_p)^2$]. Because the amount of stored energy varies with the square of the peak 60-Hz voltage, the capacitor's ability to sustain inverter operation during an outage drops off twice as fast as reductions in line voltage. As such, the power supply ride-through capability depends strongly on the capacitor being charged to full design voltage immediately prior to the power outage. This is why one 3-cycle outage may not affect electronic systems while another similar outage will shut the equipment down.

IMPACT OF VOLTAGE SAG

Utilities are permitted line voltage reductions (brownouts) to cope with seasonal demands. Large motors that accelerate high inertia loads, spot welding, and many other loads act to further drop the voltage level available at electronic power supplies in a typical facility.

Equipment shutdown and sag-induced logic errors are not the only problems. Actual damage to the DC power supply is an even greater danger. Reduced input voltage can cause excessive power supply heat dissipation, resulting in short equipment life. To maintain constant DC output as line declines, the DC-to-DC converter circuit must draw from the reservoir capacitor. Consequently, with line voltage reduced, this capacitor undergoes deep discharges between the twice-per-cycle charging periods.

Electrolytic capacitors are not designed for deep discharge, nor for the large terminal voltage variations that result. Excessive capacitor charge and discharge currents cause internal heat dissipation, producing dielectric stress and reduced mean-time between failure (MTBF). Rectifiers and DC-to-DC converter switching transistors also draw high peak currents, which raise their junction temperatures and, consequently, take a toll on semiconductor longevity.

IMPACT OF OVERVOLTAGE, SURGES, RFI, AND HARMONICS

High input voltage can puncture the power supply's rectifier and switching transistor junctions, causing MTBF degradation and breakdown. Short-term voltage surges 10% beyond nominal usually are not harmful; however, higher input voltages may overwhelm the power supply's voltage regulating capability, feeding damaging voltage levels to the electronic circuits.

Transients. High-voltage transients lasting mere microseconds can permanently wreck both the power supply and its electronic equipment load as well. Digital logic circuits that define "zeros" by voltages in the 0V-to-0.5V region, and "ones" by 4.5V-to-5V levels, are highly susceptible to inductive "kicks" that are directly impressed on their 5VDC power supply.

The power supply's reservoir capacitors ("C_1" and "C_2" in Fig. 1.11) don't absorb transient energy because their wiring inductance (negligible at 60 Hz) introduces significant isolating impedance at the MHz-equivalent frequencies of fast-rise transients. As a result, transient energy follows the line of least resistance, which is to the power supply's output terminal.

Noise. RFI and low-voltage transients created by high-current logic circuits is unlikely to damage the power supply. However, few power supply designs have careful component shielding and placement. As such, line noise can be coupled by stray capacitance to the DC output, where it disrupts communications and ITE circuits. Because this noise may be intermittent and is usually beyond the frequency range of many measuring instruments, diagnosing this source of equipment malfunction is difficult and time-consuming.

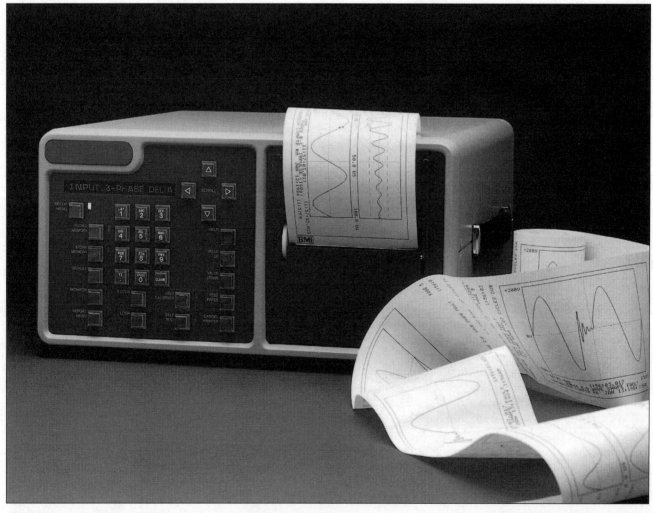
Nonlinear loads and harmonics are a special concern to persons operating systems having a large amount of solid-state electronic loads. Power monitors are often needed to help analyze disturbances.

NONLINEAR LOADS AND HARMONICS

No other problem that affects solid-state electronic equipment has been so widely discussed as harmonics and the nonlinear loads that cause it. Harmonics can cause a variety of problems, such as waveform distortion, improper voltage readings, and especially overheating in system neutral conductors. The subject, however, is often poorly understood by those who must apply corrective measures. Thus, the subject of nonlinear loads and harmonics will be discussed before all other problems that affect solid-state electronic equipment.

THE EFFECT OF LOADS

Until recently, almost all loads were linear, and those that were not were such a small portion of the total as to have little effect on system operation. Then came the solid-state electronic revolution and loads, such as computers, UPS equipment, and variable-speed motor drives, have proliferated. These electronic loads are mostly nonlinear, and they have become a large enough factor to have serious consequences in distribution systems.

Linear loads. Motors, incandescent lighting, and heating loads are linear in nature. That is, the load impedance is essentially constant regardless of the applied voltage. As seen in **Fig. 2.1**, in AC circuits the current increases proportionately as the voltage increases and decreases proportionately as the voltage decreases.

The current in these circuits is in phase with the voltage for a resistive circuit, with a power factor (PF) of unity. It lags the voltage by some phase angle for the more typical partially-inductive circuit (with a PF commonly between 0.80 and 0.95), and leads the voltage by some phase angle in a capacitive circuit. In either case, this current is always proportional to the voltage. For a sinusoidal voltage, the current is also sinusoidal.

Nonlinear loads are those in which the load current is not proportional to the instantaneous voltage (**Fig. 2.2**).

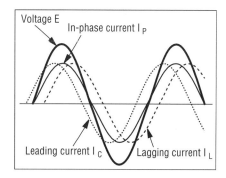

Fig. 2.1. Linear currents: I_R is a pure resistive circuit current; I_L is a partially inductive (lagging) circuit current; and I_C is a partially capacitive (leading) circuit current, which is uncommon.

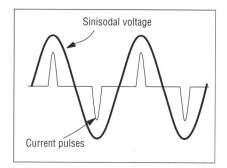

Fig. 2.2. Typical nonlinear load current.

Often, the load current is not continuous. This type of load could be switched on for only part of the cycle, as in a thyristor-controlled circuit; or pulsed, as in a controlled rectifier circuit. Nonlinear load currents are nonsinusoidal, and even when the source voltage is a clean sine wave, the nonlinear loads will distort that voltage wave, making it nonsinusoidal.

DISTORTION OF CURRENT SINE WAVES

Linear loads. Pure resistance, inductance, and capacitance are all linear. What that means is if a sine-wave voltage of a certain magnitude is placed across a circuit containing a pure resistance, the current in the circuit follows Ohm's Law: I= E/R. For a specific value of ohms, the relationship of volts and amperes is a straight line. For instance,

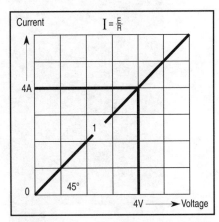

Fig. 2.3. With a linear load, the relationship between voltage and current is linear and proportional. The 45° diagonal line represents a fixed resistance.

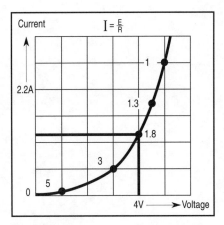

Fig. 2.4. With a nonlinear load, the load-line is curved, not straight as with a linear load. The amount of curvature is unique to each type of nonlinear semiconductor.

with 100V applied across a 10 ohm resistor, the current is 10A. If the voltage is doubled to 200V, the current is 20A, for 400V, the current is 40A, etc. This relationship is shown in **Fig. 2.3**.

<u>Inductances</u>. Ohm's Law, as it applies to inductances, is: $I = E/X_L$; where X_L = inductive reactance = $2\pi fL$ = ohms. The current is directly proportional to the voltage drop and inversely proportional to both the self-inductance and the frequency (f). If the frequency is constant (say 60 Hz), the relationship of voltage and current is a straight line, as with a resistance. Note, however, that the relationship is in magnitudes only and does not give the phase relation between E and I (E leads I by 90°). Also, the ohmic value of X_L rises in direct proportion to frequency. If frequency is doubled from 60 Hz to 120 Hz, the X_L ohms will double, etc.; maintaining a linear relationship.

<u>Capacitances</u>. A similar situation holds for capacitances. Here, the Ohm's Law equivalent is: $I = E/X_C$; where X_C = capacitive reactance = $1/(2\pi fC)$ = ohms. As with inductance, if the frequency is constant, the voltage and current relationship is a straight line one. If, however, the frequency is raised from 60 Hz to 120 Hz, the X_C ohms will be half of what is was at 60 Hz. But, the relationship is still a straight line one.

<u>Nonlinear loads</u>. Solid-state electronics is based on the use of semiconductors. These materials are totally different in that their response to voltage is not a straight line. In general, the relationship of voltage to current is represented by a curve, as shown in **Fig. 2.4**. Even this is misleading because each different solid-state device will have a response curve that is unique and different from that of other types of semiconductor-based device.

What this means is that with a nonlinear load, the relationship between the voltage and current cannot easily be predicted unless an exact curve for each device is available. With equipment containing many solid-state devices, such an approach is impossible. The only logical way is through using test instruments to plot the relationships.

The test result often is baffling. With an incoming source having a near perfect 60 Hz sine wave, the current will be shown to be significantly distorted. Mathematical (Fourier) analysis of these distorted waves, however, show that they are made up of the fundamental sine wave, plus one or more waves with a frequency that is a whole multiple of the fundamental frequency. For example: a 60 Hz fundamental and 180 Hz and 300 Hz waves when added together will result in a specific type of distorted wave. These multiples of the fundamental frequency have been termed "harmonics."

HARMONICS

Any waveshape can be reproduced exactly by adding together a series of sine waves of particular frequency, amplitude, and timing, although it may require an infinite number of them. Nonsinusoidal waveforms, therefore, consist of (and can be broken down into) some finite number of pure sine waves. How harmonics combine with the fundamental to form distorted waveforms as typified in **Fig. 2.5**. In the United States, the fundamental frequency is 60 Hz and the 2nd harmonic is 120 Hz, the 3rd harmonic is 180 Hz, the 5th harmonic is 300 Hz, the 11th harmonic is 660 Hz, and so on. The more of these harmonics present, the more the current trace will vary from a pure sine wave. The amount of distortion is determined by the frequency and amplitude of the harmonic currents.

The effect of the different harmon-

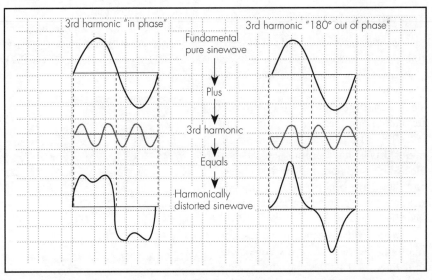

Fig. 2.5. The combination of a sinusoidal voltage waveform and a 3rd harmonic waveform creates a harmonically distorted waveform. Depending on the phase shift of the 3rd harmonic, a different signature waveform will result.

Harmonic	Sequence	Harmonic	Sequence
1	Positive	19	Positive
3	Zero	21	Zero
5	Negative	23	Negative
7	Positive	25	Positive
9	Zero	27	Zero
11	Negative	29	Negative
13	Positive	31	Positive
15	Zero	etc.	
17	Negative		

Fig. 2.6. Sequence of harmonics in a 3-phase power system. Triplen harmonics are those having a zero sequence.

ics on the operation of nonlinear loads on a power system varies.

Even-numbered harmonics. For 60-Hz power systems with nonlinear loads, even numbered harmonics (the 2nd, 4th, 6th, etc.) have been found to be considerably less likely to occur at levels detrimental to electrical systems. This is because most nonlinear loads generate odd-numbered harmonics, which are associated with a current wave shape that is a distortion of the normal 60-Hz positive and negative half cycles.

Odd-numbered harmonics. **Fig. 2.6** shows in the table the odd multiplies of the fundamental 60-Hz current and their associated sequence (positive, negative, or zero). The phase sequence of these harmonics is very important because it determines the effect of the harmonic on the operation of electrical equipment. This will be discussed later. Note that the first harmonic is actually the fundamental 60-Hz current (1×60-Hz).

<u>Positive sequence harmonics</u> (1st, 7th, 13th, 19th, etc.) consist of three phasors, each equal in magnitude, separate from each other by a 120° phase displacement and having the same phase sequence as phasors representing the normal 60-Hz current.

<u>Negative sequence harmonics</u> (5th, 11th, 17th, etc.) also consist of three phasors, each equal in magnitude, separate from each other by a 120° phase displacement; however, they have a phase sequence opposite to phasors representing the normal 60-Hz current.

<u>Zero sequence harmonics</u> consist of three phasors equal in magnitude and having a zero phase displacement from each other. Therefore, these phasors are concurrent in direction, producing an amplitude that is triple any one phasor when they combine on the neutral of an electrical system. These harmonics (3rd, 9th, 15th, etc.) are called *triplen harmonics* and are typical ones generated by phase-to-neutral nonlinear loads, such as personal computers, electronic ballasts, etc.

VOLTAGE WAVE DISTORTION

Nonlinear loads such as inverters, solid-state rectifiers used in welders, DC power supplies, variable frequency drives, and electronic ballasts for lighting, are sources of harmonics in the electrical system that feed these loads. There are specific harmonics associated with each item of equipment. Equipment manufacturers can usually provide information on the magnitude and order of harmonics generated by their equipment. However, depending on the design of the specific item of equipment, the harmonics may vary in frequency and magnitude as load changes on the equipment occur. The table in **Fig. 2.7** is a summary of the typical magnitude and order of harmonics that have been encountered with certain loads.

Ohm's Law also helps explain another phenomenon: distorted current waveshapes cause distorted voltage waveshapes in a facility's electrical distribution system. Each associated harmonic current will cause a voltage at the same harmonic to exist when it flows into that particular impedance. For example, a 5th harmonic current will produce a 5th harmonic voltage, a 7th harmonic current will produce a 7th harmonic voltage, etc.

The equation takes the form of: $E = (I \times Z)$; where Z is the impedance of an electrical load in the circuit, which in the case of motors, transformers, and similar devices is mostly inductive. Because a distorted current wave is made up of the fundamental plus one or more harmonic currents, each of these currents flowing through the impedance will cause a voltage drop across the impedance. When summed together, the result is a distorted voltage wave that mimics the current waveshape.

The magnitude of current and voltage waveform distortion in a facility's electric system will be dependent upon the relative size of the nonlinear loads with respect to that system. The distortion increases as the percentage of nonlinear loads increases.

EFFECTS OF HARMONICS ON ELECTRICAL LOADS

Motors. Positive sequence harmonics from phase-to-phase nonlinear loads will cause a 3-phase motor (either induction or synchronous) to turn in the forward direction while negative sequence harmonics will try to force motors to turn in the reverse direction. Rotation will depend upon the magnitude of the harmonics present compared to the normal 60-Hz current, and their magnitude, and there will be some torsional vibrations, which may cause serious problems.

Torque produced by motors is seriously reduced in the presence of the 5th

Load Description	Harmonic order							
	1	3	5	7	9	11	13	15
Six-pulse rectifier	100	-	17	11	-	5	3	-
Twelve-pulse rectifier	100	-	3	2	-	5	3	-
Eighteen-pulse rectifier	100	-	3	2	-	1	0.5	-
Twenty-four pulse rectifier	100	-	3	2	-	1	0.5	-
Electronic/computer	100	56	33	11	5	4	2	1
Lighting/electronic	100	18	15	8	3	2	1	0.5
Office with PCs	100	51	28	9	6	4	2	2
VFD's (range)	100	1 to 9	40 to 65	17 to 41	1 to 9	4 to 8	3 to 8	0 to 2

Fig. 2.7. Harmonic currents with typical magnitudes produced by various types of equipment. The numbers under the harmonic order are expressed in percent of the fundamental 60-Hz current.

harmonic. The 5th is a negative-sequence harmonic that produces torque that tries to turn the rotor in the opposite direction from normal. There is a torque cancelling effect, which results in the motor developing a less-than-expected ability to drive a load, drawing increased current and burning up!

As harmonic currents are drawn by loads, they act on the impedance of the source, causing harmonic distortion of the source voltage. Motors are normally linear loads, but when the supply voltage has harmonic distortion, the motors draw harmonic currents. These harmonic currents cause excessive motor heating from higher hysteresis and eddy-current losses in the motor laminations and skin effect in the windings. Thus, motors supplied from sources with voltage distorted by other nonlinear loads will also overheat unless they are derated.

Transformer (and generator) ratings are based on the heating created by load currents of an undistorted 60-Hz sine wave. When the load currents are nonlinear and have a substantial harmonic content, they cause considerably more heating than the same number of amperes of pure sine wave. There are several major reasons for this.

Hysteresis. When steel is magnetized, the minute particles known as magnetic domains reverse direction as the current alternates, and the magnetic polarity also reverses. The magnetizing of the steel is not 100% efficient, since energy is required to overcome the friction of the magnetic domains. This creates hysteresis losses, resulting in heat in the steel laminations of the core. The hysteresis losses are greater for a given rms current at the higher-frequency harmonics, where the magnetic reversals are more rapid than at the fundamental 60 Hz.

Eddy currents. Alternating magnetic fields also induce currents into the steel laminations when the changing magnetic flux cuts through a conductor. These "eddy currents" flow through the resistance of the steel, generating I^2R eddy-current heating losses. Because of the higher frequencies, eddy-current losses are considerably greater for harmonic currents than they are for the same rms value of 60-Hz current.

Eddy-current losses in magnetic materials increases significantly in the presence of the 5th and 7th harmonics. The losses that can be expected are in the order of the square of the order of the harmonic. Thus, losses for these two harmonics can be 25 to 49 times higher than normal. The effect is a significant increase in the heating of the equipment.

Skin effect. A lesser, but still considerable, heating effect at higher frequencies is caused by the "skin effect" in the conductors. Currents at higher frequencies are not distributed evenly through the cross-section of the conductor. Magnetic fields tend to force current to flow toward the outside or skin of the conductor. This effect increases as the frequency increases, and also as the magnitude of current increases. At higher frequencies, the center of the conductor carries little or no current. Therefore, the effective cross-section of the conductor is decreased, and its resistance is increased. It behaves as a smaller conductor of lower ampacity. As a result, a given current at harmonic frequencies causes more conductor heating than the same current at 60 Hz.

The result of hysteresis, eddy current, and skin effect is that the transformer or generator carrying no more than its full-rated rms current, but supplying nonlinear loads with a high harmonic content, will overheat, sometimes to the point of failure.

Conductors. Harmonic currents also can cause overheating of conductors and insulating materials as a result of skin effect. The higher the frequency, the less skin depth available in the conductor.

Capacitors. Since the impedance of a capacitor is frequency dependent (decreasing reactance with increasing frequency), capacitors will be negatively affected by harmonics. Power factor correction capacitors will appear as very low impedance paths and tend to attract harmonic currents. This results in blown fuses and bulging capacitor cases. This is usually a result of the circuit being tuned or resonating at one of the harmonic frequencies.

Overcurrent protective devices. Thermal overcurrent protective devices, such as fuses and inverse-time circuit breakers, are affected by increased skin-effect heating at the higher harmonic current levels. This excess heating can cause shifts in the devices' time-versus-current characteristics, resulting in nuisance tripping. Older circuit breaker operation depended on electromagnetic force proportional to the square of peak current, not rms current. A high 3rd harmonic current, resulting in an abnormally high overall peak current, could cause these breakers to trip prematurely. Most new circuit breakers with electronic trip devices respond only to true rms current and not peak current.

Metering. Harmonics can cause errors in induction watthour meter readings. Since the induction disks are designed to monitor nondistorted fundamental current, when induction-disk watthour meters are applied on nonlinear loads, the harmonics may cause the disk to rotate faster or slower than the same rms current at 60 Hz, depending on the specific harmonic content. If the watthour meter is used for billing, this can result in utility bills that are too high or too low; in most cases, too high.

Protective relays. When protective relays are subjected to system harmonics, relay misoperation is possible, possibly resulting in undesired pick-up values, changes in voltage and current operating characteristics, and false tripping.

Electronic equipment. When the system voltage waveform becomes distorted, electronic equipment also can malfunction. For instance, electronic clocks that count zero-crossings in the waveform may not operate correctly because the distorted waveform provides more zero-crossings than a nondistorted waveform. Thus, these clocks will run fast, causing the equipment they control to incorrectly operate.

Voltage regulators. In addition to excessive heating, harmonic currents can cause other serious problems for generators. Electronic means are used to regulate the output voltage of the generator, to control the speed of the engine or prime mover (and thus the output frequency of the generator), to parallel generators, and to share the load proportionately among the paralleled units. Many of these control devices use circuits that measure the zero crossing point of the voltage or current wave. At 60 Hz this is fine; but with a

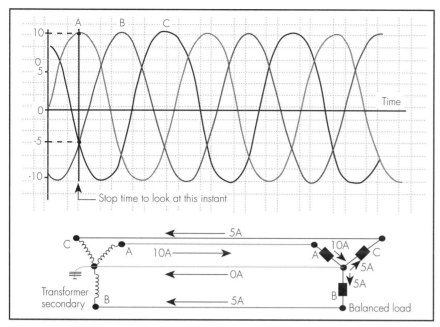

Fig. 2.8. Stopping time at any instant, the amplitudes of each of the fundamental (60 Hz) phase currents when added together equal zero. Thus, the neutral current is zero.

15th, 21st, and so on), which are the additive zero-sequence currents. These currents are generally refered to as the "triplen harmonics," and are especially troublesome. It is, however, the 3rd harmonic that has the greatest effect in causing overheated neutrals.

Other odd harmonics (5th, 7th, 11th, 13th, and so on) add in the neutral, but the total neutral-harmonic current is somewhat less than the arithmetic sum of the three harmonic phase currents. Mathematically, the total is the vector sum of the three currents. The phase angles between the three phase currents result in partial addition and partial cancellation.

The theoretical maximum neutral current with harmonics is at least 1.73 and perhaps as much as 3.0 times the phase current. (There is dispute as to the true maximum value between these limits.)

For pulsed loads, the pulses can occur in each phase at a different time. They will return in the common neutral, but they will be separated by time; therefore, there will be no cancellation. If none of the pulses overlap, the neutral current can be three times the phase current.

high harmonic content, there may be many more zero crossings than the normal ones for 60 Hz. This can cause hunting and instability in speed and frequency control, and can make the paralleling of generators difficult or impossible.

Communication systems. Cross talk caused by harmonic pickup on communication links between sensitive electronic equipment can occur, resulting in erroneous data transmission.

OVERHEATED NEUTRALS

In a 3-phase, 4-wire system, single-phase line-to-neutral load currents flow in each phase conductor and return in the common neutral conductor. The three 60-Hz phase currents are separated by 120° and for balanced 3-phase "linear" loads, they are equal. When they return in the neutral, they cancel each other out, adding up to zero at all points. Therefore, for balanced 3-phase, 60-Hz loads, neutral current is zero. This is shown in **Fig. 2.8**.

For 2nd harmonic currents separated by 120°, cancellation in the neutral is also complete, resulting in zero neutral current. This holds true for all even harmonics. This is one of the reasons even harmonics are not considered to have a significant effect upon electrical equipment and distribution systems.

For 3rd harmonic currents, the return currents from each of the three phases are in phase in the neutral (**Fig. 2.9**), and so the total 3rd harmonic neutral current is the arithmetic sum of the three individual 3rd harmonic phase currents. This is also true for all odd multiples of the 3rd harmonic (9th,

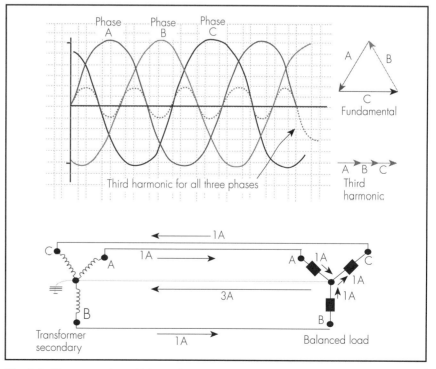

Fig. 2.9. Shown are sinusoidal waveforms of each phase current, along with that of the 3rd harmonic for all three phases. While the fundamental (60Hz) currents of the phases cancel each other, the 3rd harmonic (180 Hz) currents of each of the phases add together.

HARMONICS AND POWER FACTOR

While power factor (PF) isn't a direct measure of output-to-input efficiency, it certainly is a true measure of how well power system capacity is used. When PF is low, the power system is required to supply or carry a sizable amount of nonworking energy in order for the load to operate. When PF is improved, more work-producing energy is gotten out of the same (or sometimes smaller) power system.

The PF triangle, as shown in **Fig. 2.10**, is usually used to describe the above condition. Here, the work-producing power, which is kW, is one leg (horizontal), and the nonworking power, which is kvar, is the other leg (vertical) of the PF triangle. If these two functions are represented as vectors, then their vector sum is apparent power, or kVA. This kVA is the total capacity needed for the power system to perform the required work, which is determined by the length of the kW vector.

The angle between kW and kVA is defines the PF of the system in question. The larger the angle, the poorer the PF. Conversely, when the angle is very small, the size of the power system (in kVA) will be almost the same size as the kW of work. In this latter case, the PF will be very close to unity (1.0), resulting in an efficient handling of energy.

What causes this triangle to exist in the first place? And what can make the PF so low that system capacity almost twice that of the load is needed? Answering these questions requires a look at traditional power systems.

To power an electrical load, magnetiz-

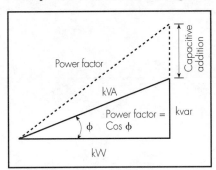

Fig. 2.10. The PF triangle consists of the work-producing power (kW), and the nonworking power (kvar). If the two functions are represented as vectors, then their vector sum is apparent power (kVA).

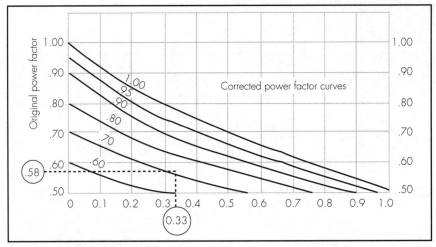

Fig. 2.11. Relationship between existing PF, corrected PF, and resulting released system capacity. By plotting a horizontal line from existing PF (0.58) to corrected PF curve (0.70), and then plotting a vertical line from this intersection down, the amount of kVA released per kW of load is seen to be 0.33.

ing energy is sent to satisfy the demands of coils, wires, etc. It prepares the appropriate electrical equipment for work, but does not actually do any work itself.

An analogy is a water piping system. Here, a certain amount of water has to be put into the pipe before water is pushed out the other end. The electrical load, which requires magnetizing energy before it can begin to do work is, in effect, asking for magnetizing energy to fill its pipeline.

Looking at Fig. 2.10 again, note that the vertical component of the PF triangle (kvar) also represents reactive energy. This energy comes in two forms: Inductive and capacitive, with the two being 180 electrical degrees opposite each other. When the inductive demands are large, the PF is low. What's happening is that the reactive requirements of the load are stealing capacity from the power system, requiring more system size to do a smaller portion of work.

Rather than oversizing the system to allow it to transport this reactive energy, capacitors can be installed, thereby offsetting some or most of the demand for inductive reactive energy. This increases the PF, permitting the kVA to be less for the same amount of work, and adding a measure of efficient power handling to the system. **Fig. 2.11** shows the relationship between existing PF, corrected PF, and the resulting released system capacity.

Many people are surprised when new equipment requiring high-frequency currents is installed. What was once a 0.88 power factor (PF) before the installation is now a 0.70 PF with the new equipment energized. How can this alarming condition be explained?

Power factor is the ratio between the true power (W) consumed and the product of the voltage and the current:

PF= W/EI,

which can be transposed to:

W= EI × PF.

For a sine wave, the power factor is often expressed as the cosine of the phase angle theta (ϕ) between the voltage and current, and this equation becomes:

W= EI Cos ϕ.

This is true *only* for a sine wave, *not* for nonlinear loads. The only accurate way to find the nonlinear power factor is to measure the average instantaneous power, and divide that by the product of the true-rms voltage and the true-rms current.

Look at **Fig. 2.12** and compare it with the PF vector diagram of Fig. 2.10. Notice that an additional vector is added called "distortion." Also note the traditional PF vector diagram is two-dimensional, while Fig. 2.12 is three-dimensional, with the kVA vector coming out of the page. This vector includes the contribution that distortion places on system capacity.

Where does this distortion come from? The high-frequency currents re-

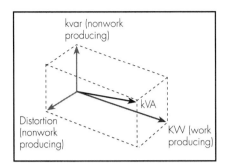

Fig. 2.12. Three-dimensional vector diagram shows that kvar and distortion are forms of nonwork-producing energy. The vector addition of these two plus the work-producing kW results in a longer kVA vector.

quired by nonlinear loads, that's where. The resulting PF is called distortion PF.

To understand the total PF picture, understand that both kvar and distortion are nonwork-producing. The combination (vector addition) of both with kW, which is work-producing, results in a longer kVA vector (having a greater magnitude). In other words, the greater the distortion and displacement PFs in the system, the larger the system capacity required to power a specific load.

HARMONIC RESONANCE

The best way to deal with harmonics is to eliminate the high-frequency currents entirely. If pure capacitance is used, that's asking for trouble because the impedance of capacitors decreases as the system frequency increases. Thus, these devices become "sinks" that attract high-frequency currents, possibly leading to their overheating, failure, or tripping off line.

Resonance occurs when the inductance in the system and the natural capacitance of the system (or added capacitors) form a tuned circuit resonant at one or more of the harmonic frequencies produced by the nonlinear loads. These resonant circuits can build up unusually high voltages, causing insulation breakdowns and equipment failures. Also, resonant circuits can draw very high currents, overloading some portions of the circuit. Since the inductance and capacitance of a system are unpredictable and often variable, resonance effects are difficult to calculate.

Resonance is becoming an increasing problem as more capacitors are being installed on facility distribution systems in order to improve power factor. Capacitor failure can result due to high harmonic content in a system. The reactance of a capacitor goes down as the frequency of the applied voltage goes up. At higher harmonic frequencies, the reactance of capacitors added to the system for power-factor correction or surge suppression can be so low as to constitute a virtual short circuit.

As seen in **Fig. 2.13**, a special concern as capacitors are added is that the 5th and 7th resonant harmonics become more predominant. Six-pulse converters used in many variable-speed drives are sources of large amounts of these harmonics (one above and one below the number of pulses). Thus, as resonance occurs, the harmonic currents flow back and forth between the source of harmonics and the capacitors. The current can be 250-300% of normal, causing the capacitors to be taken off line by their protective devices.

There is one solution to the resonance problem that works well: The use of harmonic traps (harmonic filters). While capacitors alone provide vars, protected capacitors provide both vars and high-frequency current. In these assemblies, the capacitors are sized in combination with inductance, tuned to the lowest frequency of concern, and applied in shunt as conventional pure capacitors at the load terminals. The impedance of the assembly will be low for the tuned frequency only and will, as a result, attract and dissipate those currents having that frequency.

MAJOR SOURCE OF HARMONICS

The trend of replacing mainframe computers with smaller workstation units and PCs has its good and bad points. One of the good points is that power quality issues associated with large, raised-floor installations no longer are a worry.

Although this downsizing might solve some power quality problems, they are replaced by others introduced because of new power supply technologies used. And, the problem sources are on every desk instead of being located in a specific data processing area. Whole bundles of interconnected signal wiring are present. Ground loops are ever present; noise interference through wires as well as through air (radiated interference) is common. And most importantly, harmonic interactions play havoc with existing branch circuits and feeders.

PC and workstation power supplies demand large quantities of harmonic currents in addition to 60 Hz work-pro-

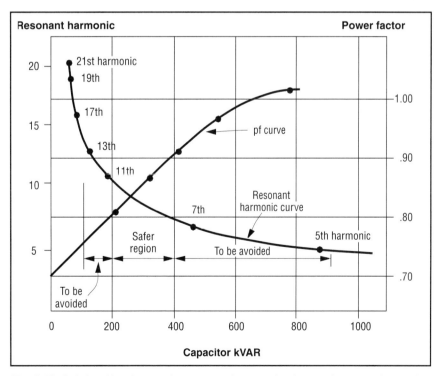

Fig. 2.13. Relationship of power-factor correction capacitor kvar and resonant harmonics.

Technician uses a multimeter to take readings while troubleshooting a VFD at its main enclosure. Two of three power transistor boards have been removed (laying on floor) to permit voltage check at internal circuitry.

ducing current. The harmonic spectrum will normally consist of 85% third harmonics, 55% fifth, 40% seventh, 25% ninth, and lesser amounts of additional odd harmonic orders. (All of these are percentages of the fundamental 60 Hz current.) When all these currents are summed up and added to the fundamental (by the square root of the sum of the squares), the total rms current, or heating value, is 50% greater than the 60 Hz current alone.

Switched-mode power supplies (switchers) in computers and similar phase-to-neutral nonlinear loads generate large 3rd harmonic current, as well as other triplen harmonic currents because they demand current only at the peak of the voltage waveform. Unless filtered from the system at the power supply, the triplen harmonic currents will flow and seek the path of least impedance; that is, through the neutral conductors towards the transformer (source of power). With high neutral current and undersized neutrals, a facility can experience excessive neutral conductor heating, resulting in possible fires, short circuits, or bus failure.

Switchers are used in most electronic equipment using solid-state "chips." These devices require DC power at between 3 and 15V. At one time this voltage was provided by a conventional linear power supply based on a transformer, rectifier, filter, and voltage regulator. Now, the switcher and controls are transistorized and switch the DC voltage on and off at a high frequency, which can range from 10 to 100 kHz. The DC output from the diodes is twice that of the switching frequency, ranging from 20 to 200 kHz. Because of the high switching frequency, the transformer and the filter can be very small and light compared to their 60-Hz equivalents. The output voltage is controlled by the switcher, eliminating the series regulator and its losses. The current pulses produced by these units are extremely nonlinear and high in harmonics.

While switchers can be the source of problems, they also have many advantages that dictate their use. Overall switcher efficiency is about 75%, as opposed to linear supply efficiency of about 50%. Losses have been reduced by half, reducing operating costs and further reducing size by requiring less cooling. Better energy storage provides longer ride-through, about 16 ms (one cycle) as compared to about 4 ms for the linear supply. The linear supply requires input voltage within 10% of nominal, while the switcher can tolerate voltage dips of about 20%.

While switchers constitute a major source of harmonic currents, there are many other sources. Depending upon the nature of the loads that predominate in a particular facility, these other sources may actually be the main contributors of harmonic currents.

Fluorescent ballasts. In fluorescent lighting fixtures with conventional magnetic ballasts, the third-harmonic current content is typically in the range of 13% to 20% of the fundamental 60-Hz frequency. Electronic ballasts generate an even higher third-harmonic component, as high as 80%.

Variable-speed drives. For power quality considerations, variable-speed drives can be divided into two basic groups: voltage source inverter (VSI) drives, and current source inverter (CSI) drives.

VSIs use large capacitors in the DC link to provide constant voltage to the inverter. The inverter then chops the DC voltage to provide the variable frequency AC voltage to the motor. VSI-type adjustable speed drives typically have a high percentage of harmonics associated with their input current, especially the 5th and 7th.

CSI drives, on the other hand, are used for larger horsepower applications where custom designs are justified. The DC link consists of a large choke to keep the DC current relatively constant. The inverter then chops this current waveform to provide the variable frequency AC signal to the motor. CSI drive currents typically have a high 5th harmonic content.

Static UPS systems. In static UPS systems, the incoming AC is rectified to DC, which is then inverted by pulsing circuits back to AC to obtain constant-frequency 60 or 415 Hz AC power. The rectifier voltage control is obtained with thyristors (SCRs). These are gated on (conducting) at any point in the cycle, turn off automatically as the voltage passes through zero, and are gated on again at the same point in each subsequent half-cycle. Output distortion of the UPS for a given load depends on the UPS design and output impedance. Thus, most UPS manufacturers specify the output voltage distortion of their equipment; 5% total harmonic distortion (THD) is typical. However, many add a disclaimer, such as, "Based on linear loads" or "For reactive and inductive loads." Such a disclaimer means that the THD figure only applies under linear load conditions.

Rectifiers. A characteristic of all rectifier circuits is that they are nonlinear and draw currents of high harmonic content from the source. Diode full-wave rectifiers are the least nonlinear, conducting as soon as the forward voltage overcomes the small (about 0.7V) forward bias required. Phase-controlled rectifiers using thyristors do not begin to conduct until gated on and are therefore more nonlinear.

Filters. Most electronic equipment uses capacitor-input filters, which are more nonlinear than reactor-input (choke-input) filters. The characteristics of the load may also increase the nonlinearity of the input current.

DEFEATING HARMONICS

Extensive use of sensitive electronic equipment has brought with it many significant advantages. Speed of response, quantum leaps in the amount of data that can be handled, accuracy, and many other advances, have changed the way the electrical industry thinks. Even such standbys as fluorescent lamps are affected; electronic ballasts depend upon digital logic.

There is a negative effect to the introduction of digital logic devices into the power distribution, control, and instrumentation. Equipment that once was considered to be invulnerable to anything but drastic voltage surges, dips, or complete power failure, is now experiencing difficulty dealing with harmonics introduced by nonlinear loads.

This chapter will focus on the various proven ways in which the problem of harmonics can be defeated.

GENERAL GUIDELINES

The following recommendations were made by the Computer and Business Equipment Manufactures Association (now ITIC) in a letter dated 1988 that included recommendations on how to overcome problems of harmonics in power distribution and utilization equipment. Most of these guidelines are still essential valid.

• Use true rms ammeters to measure load currents in all phases and neutral.

• If the steady state peak current is unknown, measure it with an oscilloscope and current probe. Measurement with moving coil or peak-hold ammeters will give erroneous information.

• Measuring instruments should have sufficient band width to provide accurate readings, taking into account the fundamental frequency and harmonic content of the parameters being measured. Current probes should be suitably rated for peak currents involved.

• Run a separate neutral to 120V outlet receptacles on each phase. Avoid using a shared neutral conductor for single phase 120V outlets on different phases.

• Where a shared neutral conductor for a 208Y/120V system must be used for multiple phases, use a neutral conductor having at least 1.73 times the ampacity of the phase conductors. A convenient way to accomplish this is to use two paralleled neutral conductors, each having the same ampacity as the phase conductors.

• Where it is not feasible to install double-size neutral conductors, protect the neutral by installing an overcurrent sensor in the common neutral which will trip an upstream circuit breaker whenever the neutral conductor is overloaded.

• Select three-phase transformers with low internal impedance, preferably in the 3% to 5% range, and always in the delta primary, wye secondary configuration. The transformer should be of three-legged core construction rather than three single-phase transformers or any open delta arrangement. An open delta transformer is not recommended because it fails to provide a low-impedance path for the third harmonic current, and does not provide the power factor correction which a delta-connected winding can supply.

• As an alternative to changing the building's main transformer and its wiring, install one or more delta/wye transformers which are designed for rectifier loads and may be placed in the office or computer room. This may be done either as part of premises wiring or with modular power centers.

• Although power factor correcting capacitors, which are properly selected and applied on the load side of the transformer, may reduce the harmonic currents the transformer may otherwise have to carry, this is not a recommended practice. Power system resonances and unexpected high harmonic current may occur unless the system has been specifically designed to avoid them. Each system reconfiguration will require re-evaluation of the design.

• A periodic inspection of the electrical system, particularly after any change in system configuration, should include measurements of phase and neutral currents, temperatures of transformers, their connections, and the connections in the total distribution system. A portable infrared temperature detector is convenient and effective for this purpose. If high temperatures are detected (above 50°C), measurements should be made to determine if excessive currents or loose connections are the cause.

k-FACTOR TRANSFORMERS FOR HARMONIC LOADS

Transformers of a building containing nonlinear loads see load currents rich in harmonic content. This causes higher eddy-current loss in transformer winding conductors and structural parts and results in higher-than-normal operating temperatures. As a result, the life expectancy of the transformer is seriously reduced.

Prior to the publishing of UL Standards 1561, *Dry-type General Purpose and Power Transformers*, and 1562, *Distribution, Dry-type - Over 600 Volts*, loading of a transformer, as part of its listing, was restricted to a sinusoidal, non-harmonic load with less than 5% harmonic content. With the publication of the standards, UL is able to list specially constructed transformers for powering harmonic-generating loads. Given the NEC requirement that electrical equipment listed by nationally recognized testing laboratories be used when available, the use of these types of transformers is now *mandated for new construction* when the load they will feed contains significant quantities of nonlinear loads. A derated transformer will *not* meet this requirement.

These special listed transformers are known as "k-factor transformers," and have a notation on their nameplates: "Suitable For Non-Sinusoidal Current Load with a k-Factor not to exceed — —." The manufacturer inserts the proper k rating.

Transformers (and motors) are usually the largest ferromagnetic loads connected to any distribution system. The load characteristics of motors tend to be more linear than transformers since motors have an air gap. Power transformers, on the other hand, typically are the first components to be seen on a distribution system. Even partial saturation of a transformer's iron core will cause a harmonic magnetizing current that differs greatly from a sinusoidal wave.

Since triplen harmonic currents (3rd, 9th, 15th, etc.) circulate in the primary winding of a delta-wye transformer, they will add to the core flux density, possibly driving it towards saturation. Also 5th, 7th, 11th, 13th, etc. harmonic currents will pass directly through a delta-wye transformer to its source of power, possibly causing overheating and damage to this source.

Calculating k-factor. ANSI/IEEE C57.110-1986, *Recommended Practice for Establishing Transformer Capability When Supplying Nonsinusoidal Load Currents*, provides a method of calculating the additional heating, or watt loss, that will occur in a transformer when supplying a load that generates a specific level of harmonic distortion. The method involves calculating the per unit (pu) RMS current for each of the harmonic frequencies. When the individual and total frequency harmonic distortions are obtained from a harmonic analyzer, the individual frequency per unit RMS currents [I_h(pu)] can be found by using the following equation.

$$I_h(pu) = \%HD(h) \div \sqrt{(THD)^2 \times (100)^2}$$

where: h = harmonic frequency order number (1, 3, 5, etc);

I_h(pu) = per unit RMS current at harmonic order h;

%HD(h) = percent harmonic distortion at harmonic order h; and

THD = total harmonic distortion.

C57.110-1986 also shows how to calculate the additional transformer watt loss as a result of a specific distribution of harmonic currents. Basically, each I_h(pu) is squared, and then multiplied by the square of the harmonic (h^2), thus obtaining $[I_h(pu)]^2 h^2$. This is done for each of the harmonic orders present on the system and then added together. This summation is called the k factor. (A typical example is shown in **Fig. 3.1**.) The k-factor is then multiplied by the eddy-current losses, which yields the amount of increased transformer heating that will occur because of the harmonic distortion. The eddy-current losses can be found by subtracting the I^2R losses from the measured impedance losses.

k-rated transformer construction. Typical k-rated transformers are 600V ventilated dry types and are available in k ratings of 4, 13, 20, 30, or 50. They typically have delta-connected primaries and wye-connected secondaries. Temperature rises of 150°C, 115°C and 80°C at rated voltage are available. Most use 220°C UL component-recog-

Transformers can be specifically designed for nonlinear loads. They have lower than normal flux densities and, thus, can tolerate a condition of high line voltage coupled with circulating harmonic currents. The "k-factor" rating of a transformer is related to its ability to support nonlinear loads. The higher the k-factor, the greater the nonlinear load it can accommodate.

nized insulation systems. Windings can be copper or aluminum.

Some units are available with six 2.5% taps, two above normal and four below normal rating. Most have electrostatic shields between the primary and secondary windings to prevent electrical noise.

To compensate for skin effect and eddy-current losses, the primary and secondary winding use smaller-than-normal, paralleled, individually insulated, and transposed conductors. Since third-harmonic current circulates in the primary delta winding, this winding is sized accordingly to limit the temperature rise to the transformer's rated value. The secondary neutral is sized to have twice the ampacity of the secondary phase conductors. The basic impulse level (BIL) is normally rated at 10kV.

DERATING EXISTING TRANSFORMERS

One way of avoiding a reduced service life for *existing* transformers carrying nonlinear loads is by derating a transformer. There are two methods of calculating the proper derating; one for phase-to-neutral loads, and the other for phase-to-phase loads.

Phase-to-neutral loads. A generally accepted method of calculating the transformer harmonic derating factor (THDF) for phase-to-neutral loads is available. This method should *not* be used to derate transformers powering phase-to-phase nonlinear loads because it is only applicable in conditions where there is a high 3rd harmonic content with high crest factor.

Harmonic (h)	Current (I_h)	I_h (pu)	I_h (pu)^{2}h^2
1	91%	.91	0.83
3	34%	.34	1.04
5	22%	.22	1.21
7	10%	.10	0.49
9	4%	.04	0.13
11	3%	.03	0.11
13	2%	.02	0.07
15	2%	.02	0.09
17	1%	.01	0.03
		Total (k-Factor) =	4.00

Fig. 3.1. Example of a typical calculation of k-factor.

Using this derating method, it is necessary to measure the true-rms phase current and the instantaneous peak phase current for each phase of the secondary. While the measurements are simple to make, the test equipment must be capable of taking these true-rms type of readings.

To determine the derating factor for the transformer, the instantaneous peak current (Ipeak) and true-rms current (Irms) measurements are taken for the three phase conductors. If the phases are not balanced, the numbers used should be an average of the three phases. These values can then be used in the following equation:

$$\text{THDF} = (1.414 \times I_{rms}) \div I_{peak}.$$

This equation generates a value between 0 and 1.0, although typically between 0.5 and 0.9. Note that if the phase currents are purely sinusoidal, the instantaneous peaks would be 1.414 times the true-rms value, and the derating factor would be 1.0 (no derating.)

The transformer rating with harmonics present would then be:

$$\text{kVA}_{derated} = \text{kVA}_{nameplate} \times \text{THDF}.$$

Once the transformer has been derated, the next step is to determine the amount of kVA the transformer is actually delivering and compare that amount to its derated capacity. The actual kVA for a balanced load can be calculated from the true-rms phase-to-phase voltage (V_{rms}) and the true-rms phase current (I_{rms}):

$$\text{kVA}_{actual} = (1.732 \times V_{rms} \times I_{rms}) \div 1000$$

If the actual kVA is close to the derated load capacity of the transformer, consider providing more cooling for the transformer. If additional cooling is not practical, loads may need to be shifted to another transformer.

Phase-to-phase loads. Performance capability of existing transformers serving 3-phase loads when subjected to harmonics can also be determined by applying the latest issue of the ANSI/IEEE C57.110 standard.

Normally, determining the extent of eddy-current loss as a result of harmonic currents would involve sophisticated computer analysis. This standard, however, can be applied through calculations made on a hand-held calculator. Although not as accurate as computer analysis, the relatively simple calculations involved are, nevertheless, conservative enough to be effective.

When direct current is passed through transformer winding conductors, a simple I^2R loss results, where R is the DC resistance of the winding. When an alternating current of the same magnitude is passed through these same conductors, an additional loss called stray loss is produced. Although all this additional loss is eddy-current loss, only that portion in the transformer windings is here called "eddy-current loss" (PEC); the portion outside the windings is termed "other stray loss" (POSL). The total load loss in watts (PLL) of a transformer can then be expressed by:

$$\text{PLL} = I^2R + \text{PEC} + \text{POSL}$$

Due to the fact that eddy-current loss is proportional to the square of the load current and the square of the frequency, nonsinusoidal load currents will cause excessive winding loss resulting in an abnormal winding temperature rise. The effect of these currents on "other stray losses" is much less critical. The I^2R loss will also increase because of these harmonic currents, although not to the extent of eddy-current loss. The basic intent of the IEEE standard is to reduce the losses in winding conductors under harmonic load conditions to that level seen under rated 60-Hz operating conditions.

The IEEE phase-to-phase loaded transformer capability calculation is based on data found in a certified transformer test report, such as listed in the appendixes of ANSI/IEEE C57.12.90, *Liquid-Immersed Distribution, Power, and Regulating Transformers,* and *Guide for Short-Circuit Testing of Distribution and Power Transformers,* or ANSI/IEEE C57.12.91, *Standard Test Code for Dry-Type Distribution and Power Transformers.* Most manufacturer test reports are based on, and utilize, these standards.

Persons needing to derate their *existing* 3-phase transformers because of an increase in nonlinear components of their essentially phase-to-phase loads should refer to the latest issue of ANSI/IEEE C57.110 for the exact calculations required.

ZIG-ZAG TRANSFORMERS

In Chapter 2, the problem of overheated neutrals was discussed and the role triplen harmonics play in causing this problem. Use of k-factor transformers or derating transformers compensate for the increased neutral currents that cause the overheating. Since normally the neutral conductor has no overload protection, triplen harmonic currents in the neutral can overheat the neutral without tripping the protective device in the circuit feeding the equipment.

A solution (besides eliminating the triplen harmonics via filtering) is the reduction of the triplen harmonics by use of a neutral current reducing transformer. The transformers used for this purpose are special low zero-sequence impedance type zig-zag transformers. They must be placed as close as possible to the source of the nonlinear loads in order to have the the maximum effect in reducing the neutral current harmonic distortion. **Fig. 3.2** illustrates how they should be applied.

The zig-zag transformer will reduce the current in the neutral by 60 to 90%, depending on the its location and the system impedance. It must be noted, however, that installation of these devices can increase single line-to-ground fault currents. They must be protected per the requirements of the NEC.

SIZING GENERATOR SETS FOR NONLINEAR LOADS

Many loads demand a better quality of supplied power, while at the same time, some loads are corrupting power with harmonics, surges, voltage dips, and other power quality problems. As a result, engine generator selection and sizing must be done with greater care than ever before.

Nonlinear loads distort the voltage waveform in a building powered by a utility service, and will have even greater impact on the output of a generator set. Most loads are relatively tolerant of voltage waveform distortion, in the sense that they will continue to operate with a total harmonic distortion (THD) in excess of 15% to 20%. However, with voltage distortion of that magnitude, problems can develop in solid-state electronic equipment.

In modern buildings, it's rare to find a situation that doesn't have electronic motor drives, UPS equipment, and other solid-state electronic devices. Many of these loads put special demands on generator sets, particularly their frequency control. Most electrical equipment in a building has been designed to work on utility power, so generator sets that don't work like a utility may cause problems.

A gen-set's generated frequency is

Sizing of modern gen-sets must take into account load concerns like harmonics, power quality, transient loads, types of motor starting, and various changing loads.

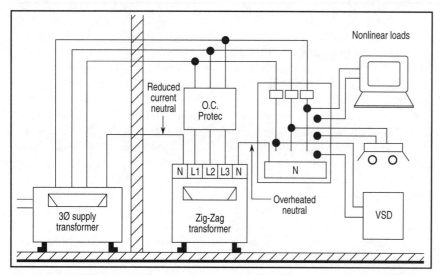

Fig. 3.2. Proper location for a zig-zag transformer used to reduce zero-sequence harmonics currents flowing because of nonlinear loads.

not as constant as utility power and will change significantly with load changes. This is the main difference between engine-generator set power and utility power, which is constant under almost all conditions. It means that frequency-sensitive loads like UPS equipment are likely to initiate alarms if variations occur when operating on gen-set power.

Look carefully at the frequency requirements of the loads being served. If they require frequency to be constant at 60 (or 50) Hz, an electronic isochronous governor for the generator set will need to be provided. For example, most UPS systems will operate over a broad range of frequency conditions to charge the UPS batteries, but will require electronic governing on the generator set to enable synchronizing output to the gen-set input power.

UPS equipment also has varying power quality requirements depending on its operating mode. When its rectifier is ramping up, relatively broad frequency and voltage swings can occur without disrupting equipment operation. However, when the bypass is enabled, both frequency and voltage must be constant, or an alarm condition will occur.

In some situations, equipment can be specified with load characteristics to minimize the transient load on a generator set, thus improving performance. For example, in many UPS control systems, the rectifier ramp rate and the maximum load drawn by the UPS can be controlled. By maximizing the ramp time and minimizing the maximum load applied, the generator set frequency will be more constant as the UPS ramps on to the generator set.

When applying UPS equipment on a generator set, put the UPS system on line last. The generator set is most stable and least affected by the nonlinear characteristics of a UPS when other loads are running on the system before the UPS load is applied.

Loads that include power factor correction or filters for power quality improvement should not be applied to a generator set operating at light load levels. The capacitive elements of these loads can cause the voltage of a generator set to rise uncontrollably at light load levels.

If the facility has a high percentage of nonlinear loads, it's desirable to use 3-phase sensing voltage-regulation equipment and a generator set with an independent power supply for the gen-set excitation system. Common examples are permanent magnet generator (PMG) and series boost type excitation systems. These make the generator set excitation system less sensitive to voltage waveform distortion.

COPING WITH GENERATOR HARMONICS

Electric generators have always been a source of harmonics because of the way these machines are designed and built. They do not generate a perfect sine wave. Depending on the pitch factor of windings and other design parameters, the magnitude and frequency of the harmonics generated will vary. For instance, the 3rd, 5th, and 7th harmonics are generated by what is considered a standard machine; that is, a generator wound with 4/5 or 5/6 pitch.

To understand this, look at **Figs. 3.3** and **3.4**. These are simplified diagrams of a full pitch stator winding and a distributed fractional 5/6 pitch winding respectively. Note that in a generator's stator core, the edges of the windings of adjacent poles ($-a_1, +a_1; -a_3, +a_4$) share the same slot. When the stator's

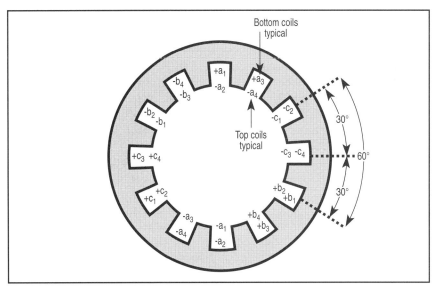

Fig. 3.3. Simplified view showing cross-section of a generator stator and the location of the windings. Shown is a 3-phase, 2-pole, full-pitch stator winding.

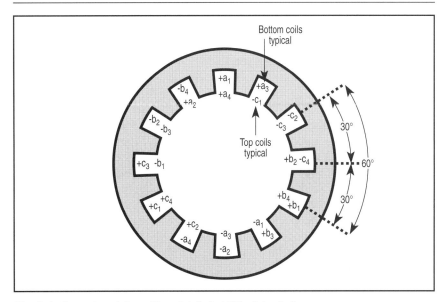

Fig. 3.4. Generator stator with a distributed 5/6 pitch winding.

slots share coils from more than one phase, as is the case here, the windings are considered to be distributed.

The individual coils in Fig. 3.3 span a full pole pitch, or 180 electrical degrees. For example, $+c_1$ and $+c_2$ are 180° from $-c_1$ and $-c_2$. Thus, the winding is a full pitch winding (6/6 pitch). In this very simplified representation of a generator stator, the coil winding can be considered to actually be two coils, one atop of another, in the same stator slot.

If all the top coils of the coil windings were shifted one slot counterclockwise, as in Fig. 3.4, then the coils would span only 5/6 of a pole pitch, or 150 electrical degrees, thus becoming a fractional pitch winding. Similar shifting by two slots yields a 2/3 pitch winding.

In addition to shortening the end connections, fractional pitch windings will decrease the harmonic content of the voltage waveform. Generators can be purchased with specially pitched windings (usually 2/3 pitch) that will not generate 3rd harmonic currents. However, the 5th and 7th harmonics generated by these machines are approximately double those generated by a standard machine.

VARIABLE-SPEED DRIVES AND HARMONICS

A variable-speed drive (VSD) is a complex piece of equipment containing many solid-state elements that can cause harmonic distortion on the power lines to which they are attached. There is no one specific solution to these problems because of the many types of VSDs availble. Thus, the need is to understand what the different types are, then see how their negative effects can be defeated.

A VSD consists of a solid-state rectifier that converts AC to DC, which is then reconverted by a solid-state inverter to a regulated AC output. It is the converter part of the VSD that dominates the interaction of the drive with its source system. Therefore, how it functions determines the extent of the harmonics it will introduce into the electrical distribution system to which it is attached.

Converter. The basic components used in the converter module are either

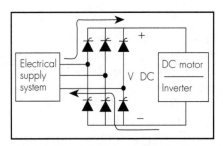

Fig. 3.5. Block diagram of a basic 3-phase, 6-pulse thyristor drive, which is the most common type used today. Arrows denote phase A current.

thyristors, transistors, or gate turn-off thyristors (GTOs). Transistors have found applications in smaller sized drives, and GTOs more commonly applied in large, specialized medium voltage applications. This leaves conventional thyristors as the most commonly applied items.

Fig. 3.5 shows a basic 3-phase, 6-pulse thyristor drive, which is the most common. Note the path for phase-A current. In a full 6-pulse converter, the thyristors operate in pairs by switching the load current among the six thyristor pairs six times per AC cycle. The thyristors shown will conduct during the positive 1/2 cycle after a gate pulse (steep voltage spike) has been received. As the current waveform crosses zero and the current tries to reverse itself through the phase-A thyristor, the negative bias causes the thyristor to turn off (commmutate). At this same time, phase-B thyristor begins to conduct, causing a brief but severe phase-to-phase short circuit.

The short circuit results in commutating notches in the voltage waveform, as shown in **Fig. 3.6**, which is a basic wave form of source line-to-line voltage, input current, and fundamental current over one cycle. The duration of this short circuit is a function of the total system inductance and the DC output current. Note the input current and voltage waveforms are no longer clean sinusoidal waveforms; the input current approximates a discontinuous square wave rich in harmonic content. The steepness, width, and area of the voltage commutating notches can result in much higher-frequency harmonic content as well as voltage spikes, both of which can result in drive control or stability problems. The inductance of the primary system (commutating reactance) plays a significant role in the notching, commutation, and harmonic generation.

Inverters. The inverter section of a drive can also impact the application. **Fig. 3.7** shows a voltage-source inverter (VSI) type VSD along with its wave form, **Fig. 3.8** shows a current-source inverter type, and **Fig. 3.9** shows a pulse-width modulated (PWM) type. Each of these drives impacts power quality, power factor, and harmonic content as a function of the line (source) inductances.

<u>VSI drives</u> control the DC voltage by using converter thyristors, which in turn control the inverter output voltage. The load draws whatever current it needs. The inverter does not use high frequency switching.

<u>CSI drives</u> control the current to the

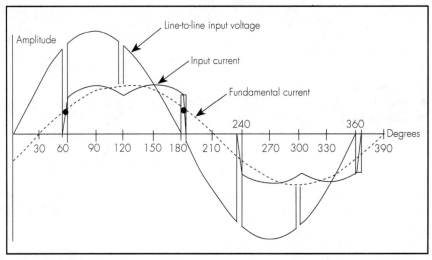

Fig. 3.6. AC input and current waveforms for a full-wave converter.

motor to maintain the required voltage and frequency with a more complex inverter section.

PWM drives use a simple diode bridge converter, which minimizes the lower order harmonics generated and does not generate commutation notches. However, this type of drive has a complex inverter section that must control both the output voltage and frequency. It uses higher frequency switching devices on the output than both VSI and CSI drives.

VSD PROBLEMS AND SOLUTIONS

Commutation notching. **Fig. 3.10** shows a typical phase-to-phase input voltage with the line notches identified. It also defines notch width and depth. The notch depth and area will differ depending were in the system they are measured (points A, B, or C in **Fig. 3.11**).

The notch depth at Point A will be 100%. At point B, it is calculated by using the following equation:

Notch depth(%) =
$[(L_1 + L_2)(L_1 + L_2 + L_3)] \times 100$

Since a line notch is a sudden change in voltage, resistor/capacitor (snubber) networks will begin to discharge/charge during commutation notches. Where multiple drives are applied on a common bus, the composite commutation notching can overwork these snubber circuits and, in severe cases, cause them to fail catastrophically. These networks are applied to absorb transient voltages occurring across the thyristors due to high-speed electronic switching. If these networks are out of tolerance or nonfunctional, a transient voltage overshoot may occur, with a high incidence of thyristor fuse blowing and/or drive component failures. Also, other solid-state electronic equipment fed from the common system can exhibit power quality problems.

Line notches also are rich in high-frequency harmonics, which are propagated throughout the power system. Solid-state electronic equipment, such as computers, PLCs, and instrumentation are especially sensitive to the high frequency content.

Solution. For this reason, it's always a good idea to keep these sensitive-type loads electrically isolated from these drives.

Line notching can also cause thyristor misfiring. This can happen when notch width exceeds gate pulse width of with excessive notch depth. VSDs should be selected to operate with notch widths of 250 microseconds and a depth of 70% of rated peak line current or less.

Solution. For those systems having notching out of limits, add commutating reactance as shown in Fig. 3.11. Generally, this is best accomplished by making the sum of $L_2 + L_3$ greater than L_1. The addition of a series commutating reactor in the drive itself or a drive isolation transformer serves this purpose.

Another benefit in adding commu-

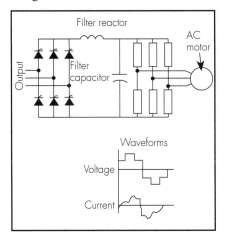

Fig. 3.7. Schematic diagram and output waveform of a voltage source inverter (VSI) type drive.

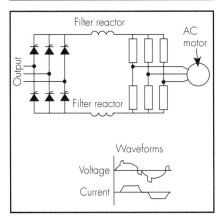

Fig. 3.8. Schematic diagram and output waveform of a current source inverter (CSI) type drive.

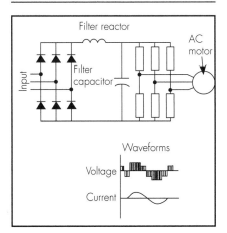

Fig. 3.9. Schematic diagram and output waveform of a pulse width modulated (PWM) type drive.

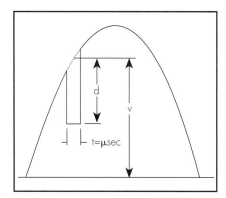

Fig. 3.10. Phase-to-phase input voltage waveform is shown, with line notches identified. The diagram also defines notch width and depth. The notch depth and area will differ depending where in the system they are measured.

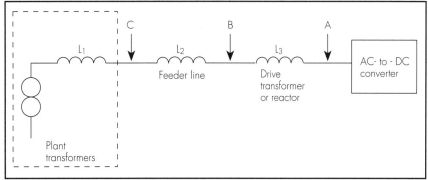

3.11. AC line impedance distribution.

tation reactance is that the current harmonics generated can be significantly reduced. However, when the addition of inductance to decrease notch depth results in too large a notch area, reducing L_1 (larger stepdown transformer) or the addition of power factor capacitors at Point C can be alternatives; they, however, affect other system design parameters such as short-circuit levels and harmonics.

Harmonics. Production of harmonics by line-commutated converters is related to the pulse number of the device. This is expressed mathematically by:

h = (np) ±1

where h = the harmonic order;
n = any integer; and
p = pulse number of converter.

For example: for a 6-pulse unit,
h = (1×6) −1=5; (1×6)+1=7;

(2×6)−1 = 11; etc. The harmonics present, thus, will be 5th, 7th, 11th, 13th, 17th, 19th, and so on. However, inaccuracies in thyristor firing, differences in thyristor characteristics, and system unbalances can cause the production of other noncharacteristic harmonic orders such as 3rd, 4th, and 6th.

VSD SOLUTIONS

Harmonic filter. One of the most often suggested method of preventing the harmonics put on the line by VSDs is the installation of a harmonic trap. This subject will be examined in depth later in this chapter.

Equipment selection. Another method is by eliminating whole categories of harmonics generated by using a 12-pulse VSD. The equation used to find the harmonics produced by a 6-pulse unit also applies here. Thus, the lowest order of harmonics are: 12-1 =11th; and 12+1 =13th. While these units are more costly than the 6-pulse types, they eliminate having to trap the 5th and 7th harmonics.

With an 18-pulse unit, the 17th and 19th are the lowest harmonics, and these do not pose any danger of resonance with capacitors used for power factor correction. This type of VSD, however, is much more costly and thus is used only in larger sized drives or for special applications.

Phase manipulation offers another method of canceling harmonics. As an example, consider the case of a 100kW 6-pulse converter. The electrical angle between firings is: 360°/6 =60°. Now consider if instead of a single unit, it consisted of two parallel 50kW 6-pulse units. If the firing position of each of the supplies adjusted so that one fires every 30° rather than 60°, an equivalent of a 12-pulse unit has electrically been created. The 5th and 7th harmonics are eliminated. Using the same technique with 12-pulse units creates a 24-pulse configuration whose lowest order of harmonics is the 23rd. A typical example of phase manipulation is shown in **Fig. 3.12**.

System alternatives. One of the techniques used in facilities having large distribution systems is the use of dual-wound transformers. The primary is delta, and secondary is wound with two output windings, a delta and a wye. The result is one transformation being 30 electrical degrees apart from the other. This automatically places the firing angle of the two parallel 6-pulse converters 30° apart, creating the equivalent of a 12-pulse unit. The same result can be achieved with using two transformers, one wired delta-delta, and the other delta-wye.

HARMONIC FILTERS

Motors and generators are particularly susceptible to harmonic distortion on the line voltage supply. As a general rule-of-thumb, the total harmonic voltage distortion should not exceed 5%. If induction motors are overheating or if unexplained burnouts have occurred, the power distribution system should be analyzed for sources of harmonic cur-

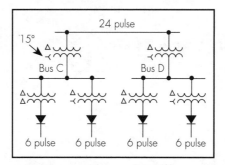

Fig. 3.12. Single-line diagram showing use of phase manipulation to cancel harmonic current.

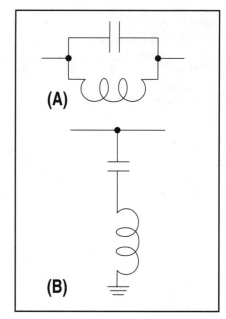

Fig. 3.13. A frequency selective harmonic filter can be used either as a high impedance series filter (A) or as a low impedance shunt filter (B).

rents. A good place to begin the search for the source of nonlinear loads affecting motors is in its controller. The use of solid-state variable-speed drives may be the culprit. If so, harmonic filters may be helpful.

A frequency selective harmonic filter can be used as either a series filter, as shown in **Fig. 3.13(A)**, or a shunt filter, as shown in **Fig. 3.13(B)**. A series filter uses a high impedance to block the power source from the flow of harmonic current being generated by the nonlinear load. A shunt filter, on the other hand, provides a low impedance path so that the harmonic current will divert to ground. Although either method of filtering will reduce harmonics, the shunt filter is usually preferred, since the series filter must be designed for full line current and insulated for line voltage.

Low-pass filters. Attenuation devices are sometimes inserted in a circuit to permit power at line frequency to pass, while attenuating transients. They are often referred to as noise filters or low-pass filters. The term "low-pass" has become more common because transient frequencies are in the kHz to MHz range, several orders of magnitude higher than the 60-Hz power.

Low-pass filters are made up of combinations of shunt capacitors and series inductances. The components must be able to withstand the high transient voltages, and the series inductances must carry the sometimes large continuous-load currents. Additionally, they must be matched to the type of transient to be attenuated and carefully matched to the input and output impedances of the circuit to be protected. Because these impedances vary, the filter must make compromises that may do more harm than good. Remember they are LC circuits and, as a result, a transient can send them into oscillation, causing the output of badly matched filters to contain more spurious voltages and frequencies than the input.

Passive filters are low-pass units that consist of one or more series-tuned inductor/capacitor elements. They provide a low-impedance path in which harmonic currents can flow; thus harmonic voltage is reduced. Passive filters are more difficult to use with auxiliary generators due to the effects that their leading load current may have on generator excitation. Also, system changes affect the filter performance.

Filters must usually be separate units installed between the source and the loads. The filters, using tuned circuits to reduce each of the troublesome harmonics, require inductors or reactors and considerable capacitance. For large loads, harmonic filters are large and heavy, but they can remove high-frequency components and leave a resultant sine wave for the distribution system.

Resonance. Interaction of devices connected to the same power bus can sometimes cause serious problems. Banks of capacitors are being routinely installed on power distribution systems in order to increase power factor. While utility billing may be reduced, the savings may be more than used up by losses due to resonance between the capacitors and equipment producing harmonics that is attached to the same bus. Of particular concern are solid-state adjustable-speed drives; their contents have high amounts of 5th and 7th harmonic currents.

The most effective means of avoid-

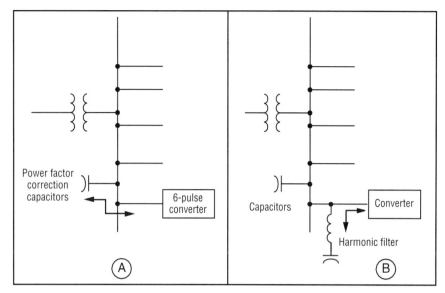

*3.14. Interaction between a load producing harmonics and power-factor correction capacitors. In (**A**), the 5th and 7th harmonic currents generated by a 6-pulse converter oscillates between the load and the capacitors. In (**B**), a harmonic trap has been installed near the converter and the harmonic currents are kept away from the power bus.*

ing high oscillating current between capacitors and harmonic-producing equipment is the installation of a filter tuned to absorb specific offending harmonics. When installed close to the line-side terminals of the equipment, the filters provide the lowest impedance path for the harmonic currents to flow. As a result, these currents will not be introduced into the rest of the distribution system, where they will cause wave distortion, and damage or tripping of power-factor correction banks. **Figs. 3.14(A)** and **(B)** illustrate this concept.

Effect of filters. As shown in **Fig. 3.15**, a typical harmonic filter (trap) is designed so that its impedance is lowest at the frequency of the harmonic that is to be eliminated. The flow of harmonic current, thus, takes place largely

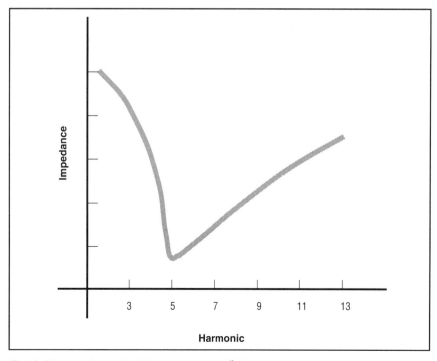

Fig. 3.15. *Impedance of a filter tuned to the 5th harmonic. Note that the filter will also be effective in reducing the 7th harmonic.*

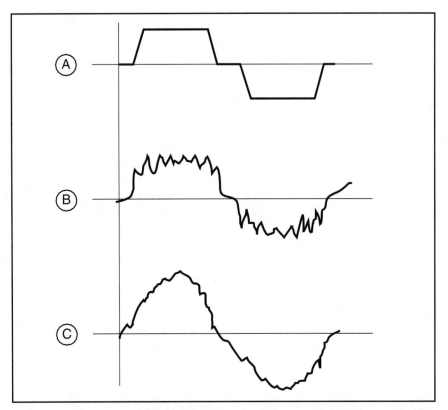

Fig. 3.16. *Filtering out harmonics with a harmonic trap causes the waveshape to change. (**A**) is a virtual square wave. By filtering out the 3rd harmonic, the waveform now looks like (**B**). When the 5th and 7th harmonics are filtered out (**C**), the waveshape conforms more closely to that of a pure sine wave.*

between the source of harmonic current and the filter. However, note that the next higher order harmonic is also significantly affected by the same filter. Thus, a filter tuned for the 5th harmonic is also helpful in eliminating the effects of the 7th harmonic.

In **Fig. 3.16(A)**, a virtual square wave, with a distortion factor of 26% is shown. When a filter is installed that reduces the 3rd harmonic (**B**), the picture changes dramatically. Although very distorted, the basic outline of a sine wave begins to be distinguishable. When the 5th and 7th harmonics are filtered out (**C**), a much closer approximation of the sine wave appears. With each succeeding harmonic that is trapped, the waveshape becomes less distorted.

Active filters. An advanced type of filter, called an active filter, senses the instantaneous sine-wave voltage at any point in the cycle. An additional power electronic converter is used to inject equal and opposite harmonic voltages and currents. It functions as an AC-to-AC filter and is technically very effective over the design range. However, the high-frequency switching devices must be selected to withstand line surges and any fault conditions to which they are to be exposed.

Active filters include clamping devices to limit the maximum voltage. Also, this filter limits the deviation from the true instantaneous sinewave voltage to 52V. When a voltage deviation is sensed, the unit switches on to provide full filtering in less than 5 nanoseconds. Not only are spikes clipped, but notches also are filled in by energy stored in the filter's capacitance. This filtering technique is relatively new and the range of ratings has not yet been established.

Active powerline conditioner. Another advanced technology approach addresses both voltage regulation and harmonic cancellation in a single integrated package. An active power line conditioner (APLC) is normally tied into the system at a panelboard and its output used to supply a cluster of sensitive electronic equipment.

Located between the power system and the loads, the APLC draws only the fundamental from the source and also acts to cancel harmonics that can be injected into the power system by the load. The APLC provides all the harmonic currents (up to the 25th) demanded by the load while maintaining voltage within close tolerances. It automatically adapts to the harmonic spectrum generated by any changes in the load or source.

An APLC can be applied without the calculations of expected harmonic currents and tuning adjustments required for tuned passive filters.

OTHER SOLUTIONS

Conventional methods of combating odd harmonic currents such as delta-wye connected transformers are not effective when the load consists primarily of switching power supplies. Delta-wye connected transformers change wave shapes but conductor heating remains the same and neutral cancellation on the wye side still does not occur.

Advanced power supplies. Power supplies can also be made to be more compatible with the present electrical distribution system. These "high power factor" supplies still draw current in short duration, high current peaks but distribute the pulses rather than aligning them with the peak voltage. Power supplies of this type are more complicated and usually have lower conversion efficiency.

Anti-phase. This approach was covered in the discussion of VFDs, but can be used in various other ways. Thus, a brief review of the technique will be repeated here. Neutral current cancellation occurs when circuit pairs with 180° phase relationship use a shared neutral. This technique, sometimes called anti-phase, causes cancellation of the fundamental and all harmonics at the shared neutral.

Anti-phase is applied by generating three additional phases 180° out of phase with the original three. The resulting six phases deliver current to the neutral in overlapping pairs (**Fig. 3.17**). When the system is balanced, neutral cancellation occurs and the resulting neutral current is zero. Cancellation occurs for the fundamental and all harmonics.

It is not necessary to completely balance the load in order for anti-phase to be useful. Indeed, in practice perfect balance does not happen. Considerable advantage is gained even though the load is not completely balanced. The important part here is that current peaks cancel instead of stacking up on each other as is the case with conventional distribution.

Conventional split-phase 240/120V service supplied to residences provides two 120V circuits in anti-phase. Conventional 3-phase, however, has a 120° phase relationship. Hence, it cannot be connected in anti-phase.

Power drawn by the anti-phase system is still in pulses. Anti-phase solves the user distribution problem but does not improve the situation for the power distributor. The power supplier will see no difference in the load whether the user is using anti-phase or not.

Inter-phase systems. Instantaneous power draw is proportional to the square of the neutral current (**Fig. 3.18**). Note that instantaneous power goes from zero to maximum six times in each power cycle. In comparison, in resistive or combination resistive/reactive loads, 3-phase power flow is continuous and uniform.

Proper power distribution and metering operation is obtained by smoothing out power flow. This can be done by shifting a second set of phases 30°. The second set of phases then have their peak current draw when the first set of phases are at zero current, smoothing power flow. The combined load looks like a resistive load to the power supplier. Customary electrical economies apply and power meters work on calibration.

A 6-phase distribution can be optimized for the power customer (anti-phase) or the power company (inter-phase) but not both at the same time. There is another power distribution sys-

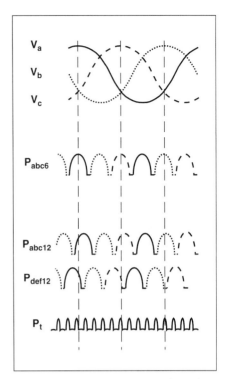

Fig. 3.18. Instantaneous power drawn from a 3-phase system goes from zero to a maximum 6 times during a cycle. Smoothed power flow can be achieved by shifting the second set of phases by 30°.

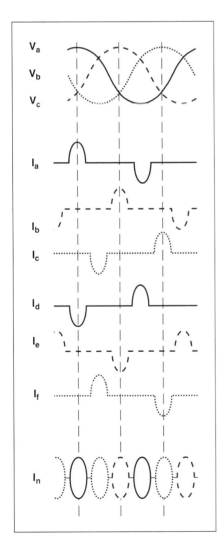

Fig. 3.17. By generating three additional phases, the anti-phase connection cancels the current in the neutral (I_n) if the loads are balanced.

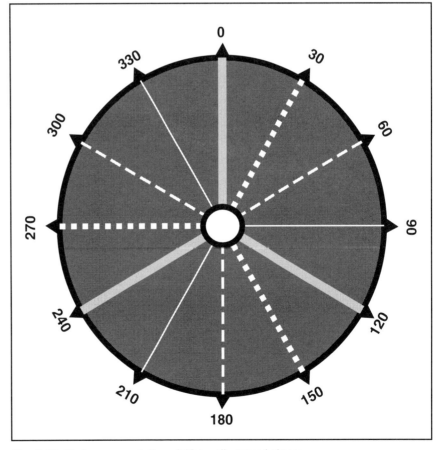

Fig. 3.19. Vector representation of 12 equally spaced phases.

tem that provides the desired features for both the user and the power distributor.

12-phase systems. A vector representation of 12 equally spaced phases is illustrated in **Fig. 3.19**. Phases at 0, 120 and 240° are conventional 3-phase power. Phases at 180, 300 and 60° are the anti-phase to them. Inter-phase to these six phases is the set of phases at 30, 90, 150, 210, 270 and 330°. Note that the phases at 90, 210 and 330° are in anti-phase to the phases at 30, 150 and 270°. The result is a smoothed power flow.

Twelve-phase is produced by a converter that receives its power as delta connected 3-phase. It produces 12 equally spaced phases in wye connection for optimum user utility. The 12 phases are distributed as 12 branch circuits.

None of the generated 12-phase circuits are in phase with the original power source. Switching power supplies operating directly from the power line draw instantaneous power peaks at a different time than any of the generated 12 phases. This adds diversity and further smoothes the power draw.

Office buildings that distribute their own power and that have 30 or more computer work stations have the most to gain from a 12-phase installation. In this case the 12-phase converter takes the place of the customary distribution transformer.

Outputs of a 12-phase converter are distributed as separate branch circuits. Branch circuit protection per the NEC is necessary for each phase or for each branch using the phase if there is more that one branch circuit per phase. Conventional 3-phase electrical distribution panels will not work. It is necessary to use a special panel set up for 12-phase input. The panel is different even though the same breakers are used and branch circuits are connected the same as in conventional 3-phase.

OTHER FACTORS AFFECTING SOLID-STATE ELECTRONIC DEVICES

The exceedingly small spacing between components in solid-state electronic equipment, especially on microprocessors, makes them very susceptible to damage from overvoltages entering either on the power supply lines or the data lines. It, therefore, is necessary to understand the nature of the problems that will affect this equipment (this chapter), before going on to considering how the equipment can best be protected (Chapter 5).

Following are the most frequent causes of disruption of proper operation of circuits involving solid-state devices.

EFFECTS OF NOISE

All microprocessors operate sequentially and are driven by an internal clock operating at millions of cycles per second. Instruction words are brought into the processor from memory and are then executed over the next few cycles. As each instruction is completed, the address of the next instruction is found in memory. If one sequential address or the instruction it contains is slightly altered, the processor is unable to continue correctly with the programmed sequence, and an error or a halt occurs. This inability to tolerate any internal scrambling of these low-voltage bits is what makes microprocessor devices so susceptible to problems from electrical noise.

Electrical noise is the enemy of microprocessor-based devices including instrumentation, monitors, displays, controllers, and computers. The manufacturers of these complex devices make it clear that providing "clean" power and grounding is important to avoid errors and malfunctions. The situation can end up so badly that equipment in the same rack set will not work properly together.

When impulse noise is observed on signal lines using an oscilloscope, it is often assumed, because the digital circuits are still working, that the noise is below the threshold that can cause problems. This is not so. Digital circuits characteristically latch in either a "high" or "low" state in which they are relatively stable. It takes a strong, deliberate signal to move a latched circuit from one state to the other. Moreover, most circuits spend most of their working life in one state or the other, and very little time in transition between states.

When a bi-stable circuit is in transition between states, however, it is very susceptible and sensitive to interference. The circuit behaves as a positive feedback amplifier and can amplify very weak noise signals to the point of saturating the semiconductor. Thus, a very weak noise signal can cause data corruption if it occurs at the moment of state transition. Although it may have a 50-50 chance of driving the circuit in the direction that was intended, it also has an equal chance of driving it in the opposite direction, causing an error.

TYPES OF NOISE

Noise signals are generally very fast and do not have the energy of a surge. For these reasons, noise usually does not cause physical damage to equipment as surges do. Nevertheless, noise can cause problems ranging from mildly annoying to very serious. Noise in the power lines causes many, if not most, of the problems that result in signal errors.

Basically, noise can be defined as unwanted electrical signals that are superimposed on a useful waveform. As such, noise can be understood as a repetitive voltage or frequency transient. Essentially, there are two different types of noise: normal-mode (also called transverse-mode) noise; or common-mode noise.

In practical applications, the noise in a given circuit contains a component of both normal-mode noise seen in **Fig. 4.1(A),** and common-mode noise seen in **Fig. 4.1(B).**

Normal-mode noise (transverse-mode) is the voltage noise that always exists between a pair of conductors. In

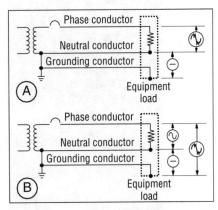

Fig. 4.1. Normal-mode transient (A) appears between the power conductors, while common-mode transients (B) appears between each power conductor and ground.

a 3-phase connection, normal-mode noise is a noise voltage that appears equally for each phase when an instrument is connected between the phase conductor and the grounded conductor. This type of noise and its origins will be discussed later in this chapter.

COMMON-MODE NOISE

By definition, common mode noise is an electrical signal that occurs on all conductors of an electrical circuit at the same instant. It's illusive since, when you measure line-to-line or line-to-neutral voltages, there will be no voltage difference at any point in time. But, while reading zero volts on a voltmeter of other measuring device, there may be significant voltage between a conductor or neutral and the safety ground connection. Another term used to describe common-mode noise is N-G (neutral-to-ground) voltage.

One effective way to "see" this noise is to measure between the power system neutral in a panelboard and the case ground (signal reference) of a load device having a digital signal system. A circuit where the power system neutral and the equipment case ground are not referenced to the same potential may experience very sharp surges or spikes similar to the steep waveshape shown in **Fig. 4.2**.

Common-mode noise (N-G voltage) is a major problem that can destroy the integrity of electronic equipment low-level signals. These voltages drive currents through the multitude of grounding paths connecting raceways, equipment, and building steel, and through the solid-state electronic circuitry.

It is hard to understand how there can be any measurable voltage on N-G terminals when they are tied to a common point at the service entrance panel. The answer lies in the definition of the two categories of conductors.

- Grounded (neutral) conductors are current-carrying wires from the building service panel or separately derived power source to the load equipment.
- Equipment grounding conductors are noncurrent carrying conductors whose function is to carry fault currents only.

Although they are bonded at the service, the fact that the neutral is carrying current while the equipment grounding conductor usually (except under fault conditions) does not, accounts for the difference in potential. The neutral conductor as a current-carrying wire has impedance and, as such, current flow creates an IR drop within the wire, which is measurable when a voltmeter is connected between the service panelboard and the load.

With no current flowing in the ground wire, it will be at ground potential relative to the common point at the panelboard whether voltage is measured at the panelboard or at the load. As a result, the difference between the voltage drop in the neutral between load and the panelboard, and no voltage drop in the grounded wire will be seen as a voltage drop between neutral and ground at the load.

For example, if a load is supplied by a 120V panelboard, there might only be 118V at the load. The 2V drop would be due to IR losses in the line and neutral. In this case, when measured with a voltmeter, the voltages are: 120V line-to-ground at the panelboard, 119V line-to-ground at the load; 1V neutral-to-ground at the load and, 0V neutral-to-ground at the panelboard.

Because N-G voltage is caused by current on the neutral conductor, it can be calculated by Ohms Law if the current is known. On the load side of the service panel, the neutral wire carries the return current from the load and, therefore, has a voltage to ground that depends upon where the reading between them is taken along the wire.

The grounding wire, as noted, normally is not a current-carrying conduc-

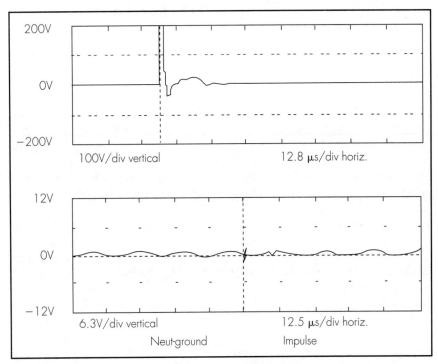

Fig. 4.2. An effective way to see electrical noise is to measure between the power system neutral in a panelboard and the case ground (signal reference) of a load device having a digital signal system. A system where the power system neutral and the equipment case ground are not referenced to the same potential may experience very sharp surges of spikes as shown.

tor and, therefore, does not normally develop a voltage. Load equipment that utilizes surge protection devices, however, may produce some leakage current on the grounding conductor, but that current will not be significant compared to that on the neutral wire.

GROUND LOOPS

While a neutral-to-ground or neutral-to-signal-reference measurement of zero volts is expected, it can actually be several hundred volts. This indicates the grounding of a device at two of more locations, creating a ground loop. One ground point may be at a different potential with respect to the others. With some impedance between the two points, there would be a potential driving a current around the loop. **Fig. 4.3** shows how such a loop might appear. This phenomena is one of the leading causes of signal driver/receiver disturbance and destruction.

Ground loops also exist within a rack set because electronic return paths are shared. All too often, provision is not made for these return currents, so they must flow on the only paths available: rack frame steel, bolted and painted rack-to-rack joints, safety-ground wires, wireway and conduit surfaces, and anything else that is available. These chance connections are orders of magnitude worse (have a higher impedance) than any rational intentional path would be, and the inevitable system return currents develop volts of mutually interfering noise over the high impedance of these chance returns. Thus, ground loops are bothersome because the path in which they flow is poor.

NORMAL-MODE NOISE

Fig. 4.4 illustrates a typical waveform with normal-mode noise. This type of electrical noise is a distortion of the normal sine wave, usually caused by power electronic circuits, arcing loads such as welders, and switching power supplies

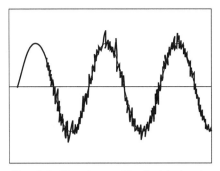

Fig. 4.4. *Waveform with electrical noise distortion.*

associated with solid-state electronic devices. Basically, this noise can be defined as unwanted electrical signals superimposed on a useful waveform. As such, this noise can be understood as a repetitive voltage or frequency transient.

A transient voltage is a high-amplitude, short-duration pulse or surge superimposed on the normal voltage. It results from the rapid release of energy stored in the inductance and capacitance of an electrical system or in a charged cloud (lightning).

There is no precise standard definition of a transient in an electrical system. Transients vary widely in current and voltage waveshapes and magnitudes. Typical peak voltages are from about twice the rms supply voltage up to thousands of volts. Durations can range from 0.5 microseconds to perhaps 200 microseconds for a single spike, and up to one cycle (16.7 ms) or even more for an oscillatory transient. Oscillatory transient frequencies may vary from a few hundred Hz up to many MHz.

Depending on the relative polarity of the power and the transient, the voltage can add to the 60-Hz sinewave voltage (a surge or spike) or subtract from it (a notch or dropout). Voltage rise can be thousands of volts per microsecond.

TYPES OF TRANSIENTS

Transients are categorized as either an impulse or an oscillatory type.

An impulse transient has a fast rise time, fairly rapid decay, and high energy content (rising to hundreds and even thousands of volts) and is unipolar. Duration can be from a few microseconds up to 200 microseconds.

A typical impulse transient is shown in **Fig. 4.5(A)**. Its impulse magnitude

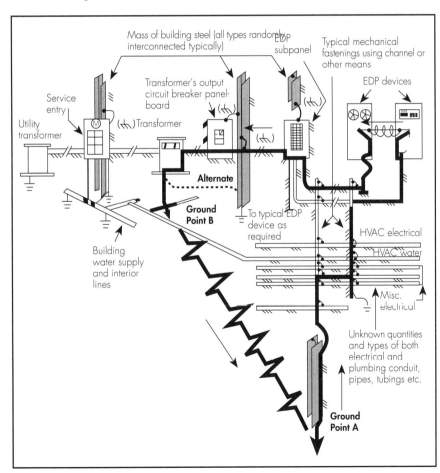

Fig. 4.3. *Example of a ground loop. Here, ground point A may have a different potential with respect to ground point B. With some impedance between the two points, there is a potential driving a current around the loop.*

is measured from the point it occurs on the sine wave and not from zero voltage, and it is commonly called a "spike" if it adds to the sine wave, or a "notch" if it subtracts from the sine wave.

An oscillatory (ringwave) transient has a fast rise time, oscillations that decay exponentially, and a lower energy content than an impulse transient (250V to 2500V). A typical oscillatory transient [**Fig. 4.5(B)**], which can last up to one cycle (16.7 ms) or even more, can have frequencies from a few hundred Hz up to many MHz.

Six different waveforms developed by IEEE, and incorporated into IEEE Standard 587-1980 help understand the characteristics of transient disturbances. These waveforms approximate the majority of disturbances, but they are not intended to be worst-case conditions—a difficult parameter to predict. This set of waveforms aid in measuring and testing transient suppression components and systems in AC power circuits with rated voltages of up to 277V line-to-ground. The waveforms can be duplicated for transient analysis by using the source impedances and short-circuit current values. The waveforms are shown in **Figs. 4.6, 4.7** and **4.8**.

In the IEEE standard, the AC power distribution system is divided into three location categories: A, B, and C according to its physical distance from the utility power distribution lines and/or its outdoor location.

Category C includes the service drop from the utility pole to the building service entrance, a run between the meter pan and main distribution panel, or overhead lines to detached buildings.

Category B includes main and subdistribution panels, industrial bus and feeder system, commercial lighting branch circuits and high-wattage utilization equipment.

Category A includes receptacle outlets more than 30 ft from a B location or 60 ft from a C location. (Note: Many homes and small commercial buildings have outlets that fall into Category B,

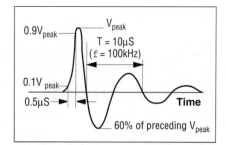

Fig. 4.6. Category A location transient is a 100kHz oscillatory wave with an open-circuit voltage of 6kV and a short-circuit current of 200A.

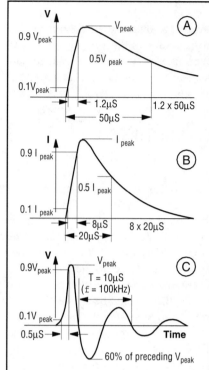

Fig. 4.7. Category B location impulse transient consists of the 6kV open-circuit voltage waveform (**A**) and the 3kA discharge (short-circuit) current waveform (**B**). The 100kHz oscillatory wave (**C**) has an open-circuit voltage of 6kV and a short-circuit current of 500A.

not Category A, because of their close proximity to the main distribution panel).

In going from the C to the A category, the transient voltage and current levels become progressively smaller be-

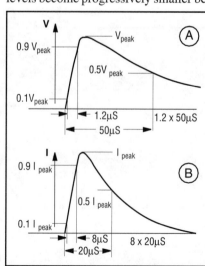

Fig. 4.8. Category C location impulse transient consists of a 10kV open-circuit voltage and a 10kA discharge (short-circuit) current waveform.

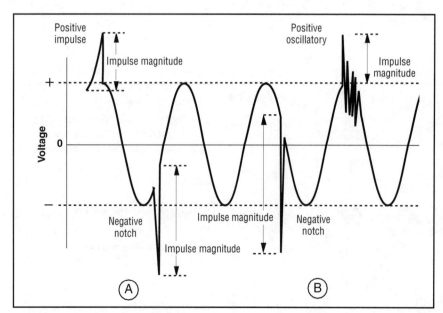

Fig. 4.5. An impulse transient (**A**) and an oscillatory transient (**B**) occurring on an AC sine wave. The magnitude of an impulse transient is measured from the point it occurs on the sine wave, not from the zero voltage or crossover point on a waveform. Thus, a transient impulse with a magnitude of 500V occurring at the top of a 120V sinewave has a peak instantaneous voltage of 670V (170V being the 120V waveform peak).

cause of the inherent impedance of a building's power distribution systems.

These waveforms and guidelines are now incorporated into the latest issues of ANSI/IEEE C62.41 standard, which is part of UL Standard 1449 Revision. The UL standard is used in testing transient voltage surge suppressors for listing.

TRANSIENTS CAUSED BY LIGHTNING AND SWITCHING

The two principal sources of transients are lightning and switching of electrical circuits.

Lightning. Often thought to be the principal cause of most surges because it is known to strike overhead power lines, lightning causes problems in facilities containing solid-state electronic equipment. While only partly true, it must be protected against. The extent of the provisions that must be made depend to a great extent to the location of the facility. **Fig. 4.9** is an isokeraunic map that shows the mean annual number of days with thunderstorms in the U.S.

A lightning stroke, which averages about 25,000A to 30,000,000V, produces high currents in lines that take a direct hit (**Fig. 4.10**). The current in these cases is limited only by the lightning voltage and the impedance looking into the system.

As shown in **Fig. 4.11**, the travel-

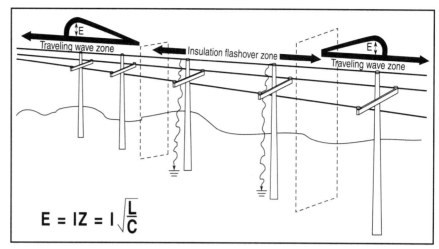

Fig. 4.10. *Traveling-wave voltage on a power line is equal to the current magnitude multiplied by the surge impedance of the line.*

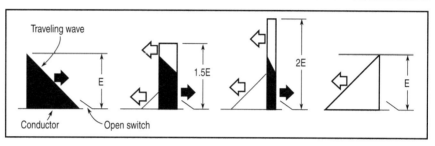

Fig. 4.11. *Diagrams show how voltage doubles when traveling wave is reflected at high impedance at far end of a cable.*

ing wave moves through the cable at a velocity of about 500 ft/millisecond and will reflect positively if the far end of the cable presents a high impedance (such as a transformer or open switch). A characteristic of traveling wave phenomena is the instantaneous voltage nearly doubles during reflection, and it is this voltage that can endanger connected equipment.

But direct strikes do not account for all the disturbances on overhead lines. What is frequently overlooked is the high electromagnetic field produced by lightning. This field oscillates at a very high frequency. Thus, a lightning stroke that misses an overhead line and strikes some nearby object or the earth induces high currents and voltages into the overhead distribution system. The induced voltage depends on the time-rate-of-change of the flux through the circuit per Faraday's Law: $e = di/dt$.

Thus, lightning produces extremely powerful, short-duration transients, either by a direct strike or a near miss. Maximum lightning voltages on unprotected indoor low-voltage systems are proportional to "residual" voltages on the primary power system, when primary arresters operate. Lightning on power lines

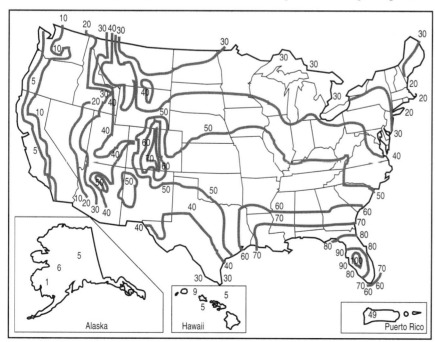

Fig. 4.9. *Isokeraunic map shows mean annual number of days with thunderstorms in the United States. The highest frequency is in south-central Florida.*

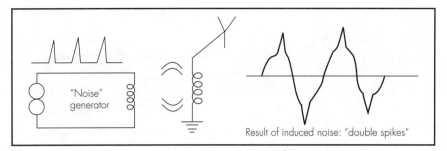

Fig. 4.12. *Noise source (noise generator) shows only the positive spikes, coupling these impulses into the nearby wire, which behaves like a coil. These coupled impulses and the reaction of the wire (coil) now do not oppose each other.*

is diverted to ground by arresters, "sparking over" to conduct the build-up of energy before damage occurs.

The topic of lightning and surge protection is discussed in greater detail in Chapter 7 of this book.

Switching. Circuit breakers protecting the lines will trip and then may reclose. The resulting voltage sags and outages can cause more problems for solid-state electronic loads than voltage transients.

Switching of utility or onsite loads also causes transients of a magnitude depending on the instantaneous rate of change of the current (di/dt). In typical inductive systems, the voltage developed across an inductance (L) is:

E = L di/dt.

Switching capacitors can cause severe transient voltages, and even normal system stray capacitance can cause transient oscillations.

RFI/EMI NOISE

Noise interference termed EMI (electromagnetic interference) or RFI (radio frequency interference) can be impressed on power lines as well as data and communication lines. They are both a form of electromagnetic interference or "noise" that can cause errors in, or loss of, transmitted data. The interference can come from automobile ignition systems, mobile radios, electrical power transmission, and the noise can also accompany high-voltage surges travelling along power lines. Low level noise may not cause failure of computer circuits but can cause upsets in the software program and alter or destroy stored data.

A high speed spike traveling down a wire into a coil generates an inductive "kick back" that opposes the amplitude of the spike, making it lower. In other words, these two forces are in series, thus canceling of reducing the effect of the spike.

Look, however, at the left of **Fig. 4.12**. Here a noise source, which in this example involves only positive spikes, couples the impulses it is carrying into a nearby conductor (center). This nearby wire behaves like a coil. The coupled impulses and the reaction of the wire do not oppose each other. Instead, they show themselves as separate spikes on the wire. In other words, there are now twice as many surges because there is no canceling (right). This is no longer a linear relationship per Ohm's Law. Now the voltage through the nearby conductor is, as before: E= L di/dt.

Note that the time (t) is in the denominator; thus, as the speed of data signals increases, di/dt gets smaller and E gets larger. As time is reduced from milliseconds to microseconds, voltage can increase to several thousand volts. Thus, induced noise is even more powerful and potentially damaging than wired noise.

The source of RF interference can also come from local pickup of signals that depend on frequency.

10 kHz to 1 MHz. This is where the real "radio frequency" (RF) range starts. AM radio stations operate in this range, and a powerful station that is close enough can cause local interference. "Walkie talkie" radios are also active in the lower frequencies. The band of frequencies between 10 kHz and 1 MHz is a transitional range, for a number of reasons. Lengths of wire become better transmitting and receiving antennas, so noise pulses can jump from one circuit to another. At 1 MHz a wavelength is 936 ft long in free space. A conductor as short as 1/20 of this length is generally considered long enough to cause pickup problems, so cables in a facility that are as short as 50 ft in length can interact with each other; when they do, they act unpredictably.

1 MHz to 10 MHz. In this frequency range, all hope is lost of reliably maintaining equal ground potential between circuits, even when associated equipment is no further away than an adjoining rack. A full wavelength can be as little as 100 ft in free space, and every wire is a potential transmitting or receiving antenna. Circuit operation becomes virtually impossible to manage as the surroundings become a large part of any circuit. Only coaxial or well-balanced shielded twisted-pair circuits are predictable.

Over 10 MHz. It is very easy to experience resonant phenomena when signals are at 10 MHz or greater because a full wavelength is less than 100 ft long. Electromagnetic coupling from one circuit to another is easy, and shielding is difficult. Wires less than 25 ft in length unexpectedly become a quarter-wave long. Thus the impedance can vary from low to high in a few feet. The voltage will be at a maximum where the impedance is highest; and when the impedance is transformed to the lowest value a quarter-wave away, the current will be at its maximum. This quarter-wave transformer effect will isolate one end of the wire from the other at the resonant frequency. Also, the high-current (low-impedance) end of the wire is the best transmitting antenna at resonance, which could be significant if it is carrying any power. A coax cable grounded at both ends will limit electromagnetic radiation if the shield is not broken.

POWER DISTURBANCES

Power-related down time of computers, according to available statistics, averages only 4.8% of the total lost time. However, there are failures in which the supply of power definitely is to blame. Input power requirements provided in manufacturers' specifica-

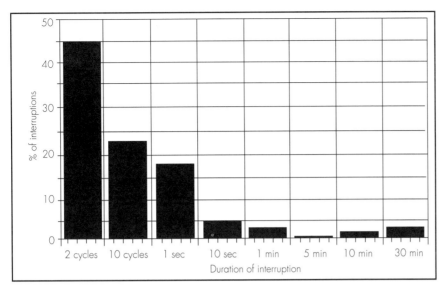

Fig. 4.13. A portion of data obtained from the Distribution Power Quality Monitoring Project survey conductied by EPRI. This survey produced information on voltage sags and swells, transient disturbances, momentary disruptions, and harmonic resonance conditions.

tions cannot be met all the time by even the best utility power service. A short duration sag in voltage from a power-system fault (short circuit) is a common disturbance causing lengthy downtime.

Just what quality can be expected from the power source? ANSI standard C84.1 lists two steady-state voltage regulation limits, Range A (+4%, -10%) and Range B (+6%, -13%). Range B should be satisfactory for solid-state equipment and will usually be met under load variations or even brownouts. Excessive phase displacement or harmonic content are rarely found in utility power. Problems of this type are usually the result of the user's own equipment.

Although the term "power quality" is not often associated with higher voltage levels, the concept is just as important for medium- and high-voltage systems as it is for 600V systems. In fact, electric utilities and their research consortiums have devoted time and funding to improve service technology at the primary voltage level. The end result has become known as "custom power."

A recent study, the "Distribution Power Quality Monitoring Project," conducted by the Electric Power Research Institute (EPRI) with the cooperation of 24 utilities, involved the monitoring of 277 sites. It produced information on voltage sags and swells, transient disturbances, momentary disruptions, and harmonic resonance conditions. **Fig. 4.13**

shows a portion of the data.

There's a rather large percentage of sags and momentary outages (the data found at time durations of 1 sec or less). These short-term power system disruptions weren't as troublesome when analog devices were the norm for end users. Now, almost every major voltage sag and momentary breaker operation causes loss of system control or a complete production shutdown.

LONG-TIME VOLTAGE ABERRATIONS

Distinct categories of recurring power disturbances identified with respect to their effect on electronic equipment, resulted in the definition of three types of disturbance (table in **Fig. 4.14**). Note that it is the duration of the disturbance, not the magnitude of the voltage, that determines the type.

Long-time voltage aberration (Type III) is an undervoltage condition that has a duration of 15 minutes or more. Usually, such aberrations are caused by one of three basic conditions:
• line droop;
• blackout; or
• brownout.

Line droop, the most common reason for long-term voltage aberrations, is caused by a reduction in the voltage transmitted over a conductor as a result of the resistance of the conductor. Obviously, the longer the conductor, the

Definition	Type I Transient and oscillatory overvoltage	Type II Momentary undervoltage or overvoltage	Type III outage
Causes	Lightning: power network switching (particularly large capacitors or inductors); operation of on-site loads	Faults on power system; large load changes; utility equipment malfunctions; on-site load changes	Faults on power system; unacceptable load changes; utility or on-site equipment malfunctions
Threshold level*	200 to 400% rated rms voltage or higher (peak instantaneous above or below rated rms)	Below 80 to 85% and above 110% of rated rms voltage	Below 80 to 85% rated rms voltage
Duration	Spikes 0.5 to 200 microsecs wide and oscillatory up to 16.7 millisecs at frequencies of 0.2 to 5 kHz and higher	From 4 to 60 cycles, depending on type of power system and on-site distribution	From 2 to 60 sec if correction is automatic; unlimited if manual

Duration (cycles of 60-Hz wave): 0 — 0.5 — 120

Fig. 4.14. Types of power-line disturbances and typical characteristics.
*denotes appoximate limits beyond which disturbance is considered to be harmful to computer.

greater the line droop.

In an effort to compensate for this reduction in voltage, the utility providing the commercial AC line power establishes a base voltage, a maximum voltage, and a minimum voltage. The base voltage refers to the desired load-center voltage; the maximum voltage refers to the maximum voltage that will be allowed for any customer at peak load; and the minimum voltage refers to the minimum voltage that will be allowed for any customer at peak load.

With these values established, voltage regulators and capacitors are arranged on the line to maintain the voltage levels as close to the base value as possible. As a result, the users close to the transformer (or distribution station) consistently experience an abnormally high voltage. Users at the end of the line consistently experience a somewhat lower voltage (because of the voltage drop over the distance). And only those in the middle of the distribution line experience the true rated voltage.

Blackouts, the second type of long-time voltage aberration, occur when the line voltage goes to zero and remains there. Such conditions can be caused by lines going down in a storm, the utility's distribution transformers failing, a fault condition causing a circuit breaker or fuse to trip, or any one of a number of other conditions.

While extended blackouts are fairly uncommon, the impact of a blackout on a given piece of electrical or electronic equipment can be devastating. This is because a blackout is characterized by the temporary loss of voltage, followed by a brief period where the voltage comes back up, and then a total loss of voltage. This fluctuation on the AC line can cause severe damage to any number of critical loads. In the case of a computer, even a momentary blackout can result in extended downtime. This is due to a number of factors, including restart of the computer and reloading of data. With the lack of redundancy in feeders to most consumers of commercial AC power, blackouts will continue to be a potentially dangerous voltage aberration.

Brownouts are a condition where the utility reduces the available voltage on the commercial AC line in a given area. On some occasions, when the power company is reaching the limits of its capacity to produce power, there will be an intentional brownout to reduce the power demand in a particular area. Unintentional brownouts, caused by a failure of some equipment at the

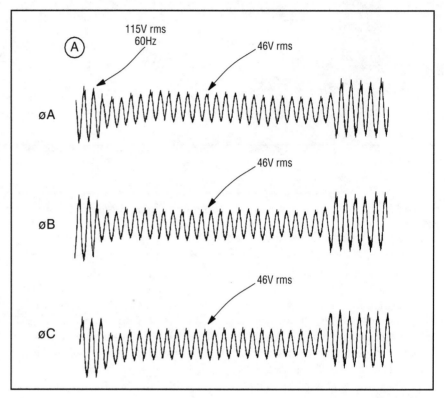

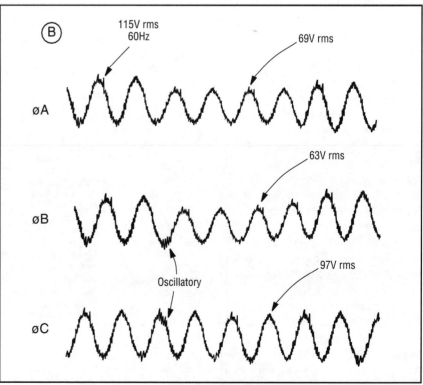

Fig. 4.15. *Monitors recorded these examples of Type II power-line anomalies, characterized by periods of undervoltage and oscillatory disturbances. Group (**A**) show a reduction in voltage on all three phases caused by a lightning fault, and (**B**) undervoltage with surerimposed spike and oscillatory disturbances recorded during a thunderstorm.*

utility's distribution points, are also quite common.

Typically, most brownouts go unnoticed. This is because the only evidence of such an event is a slight dimming of the lights in the area. Yet brownouts are quite common. While some loads require constant amounts of power, regardless of the voltage, and will continue to operate as long as it is within an acceptable range, brownouts can have a significant impact on many other types of electrical and electronic devices. Equipment such as an induction motor will run considerably hotter under these low-voltage conditions. As a result, there is a tendency for such devices to fail prematurely.

INTERMEDIATE VOLTAGE ABERRATIONS

Intermediate voltage aberration (Type II) is an overvoltage or undervoltage that has a duration of a few cycles to a few minutes (**Fig. 4.15**). Typically, this type of voltage aberration can be caused by a user on the line starting or stopping a large electrical motor, the power company changing transformer taps, or a fault condition on the line. Load-related intermediate voltage aberrations are by far the most common. Such events can result in both undervoltage and overvoltage conditions on the line.

For example, when a large electrical or electronic device is started—say a large motor, an elevator, or an air compressor—it may require an initially large surge of power. During that period of time, the line voltage sags for several cycles and then returns to normal.

Similarly, when that same large load is later turned off, it can result in an intermediate overvoltage condition on the AC power line. When a large load is suddenly removed from the line, the power-line inductance will try to maintain a constant current flow by generating a transient voltage. This voltage increase may, depending on proximity, be of sufficient magnitude to damage some types of equipment.

The switching process involved when the utility company changes transformer taps to vary the output voltage can result in an intermediate undervoltage or overvoltage condition until the circuit is able to stabilize.

When a fault condition is present on the commercial AC line, there usually is an unwanted, abnormally heavy load on the line. It can be caused by a simple short circuit, a downed power line, or a failure in a building's wiring. The line voltage will tend to drop almost to zero until the fault clears. At that point, the voltage rises, and due to the inductance of the power line, a transient voltage increase takes place. This rapid fluctuation of the line voltage can prove particularly damaging to many types of unprotected electrical and electronic equipment.

As described in Chapter 1, ITIC (CBEMA) has standardized on voltage limits within which computers and similar electronic equipment manufactured by its members should operate reliably. Voltages outside these limits could cause erratic or incorrect operation.

Actual site voltage measurements of incoming utility power show that the utilities, in spite of their best efforts, often cannot maintain the specified voltage tolerances at all times. Utilities show numerous brief instances of both above-tolerance and below-tolerance voltage deviations when monitored by equipment capable of recording short-time excursions.

SHORT-TIME VOLTAGE ABERRATIONS

Short-time voltage aberration (Type I) is known as a transient, surge, or noise, and usually has a duration of a few cycles or less. A transient voltage is a short-duration pulse or surge superimposed on the normal voltage, as described previously. It results from the rapid release of energy stored in the inductance and capacitance of an electrical system.

Damage from transients can be immediate. The transient voltage, in this case, causes a breakdown of solid-state components due to the energy of the transient. Damage also can be latent. This is a much more subtle and mysterious effect, where insulation or components are severely stressed by a single or repeated transients, but not to the point of immediate failure. Each hit weakens the ability to withstand additional stress. At some later time, a transient or other stress (that would normally cause no problem) occurs at the weakened component, and it fails unexpectedly without apparent cause.

FREQUENCY ABERRATIONS

Frequency aberrations occur when the line frequency maintained by the power company (usually between 59.98 and 60.02 Hz in the U.S.) varies from this precise window.

To achieve the desired frequency over the course of a given period, the utility could generate a lower frequency for a given period and then catch up by increasing the frequency above 60 Hz for a short interval. While this results in generation of an *average* of 60 Hz for the period, the resulting variations can have a negative impact on loads.

On occasion, the power company will, of necessity, switch distribution paths, so an area normally supplied by substation A is temporarily supplied by substation B. If the two substations are not precisely in phase when this switch occurs, it can result in switching transients that show up as a momentary frequency aberration.

More commonly, frequency aberrations are a problem within facilities where secondary power services, such as diesel-engine generators or gas turbines, are used. Usually found in applications where an emergency power source is critical, auxiliary power sources can be very difficult to maintain at a constant 60-Hz frequency. As a result, electrical and electronic devices operating on auxiliary power sources can be disrupted by high, low, or varying frequency.

INPUT POWER GUIDELINES

The precise extent to which solid-state equipment is susceptible to these disturbances is difficult to determine. Nevertheless, some guidelines are available for built-in protection against line disturbances and to aid in the diagnosing of power-related failures. The generalized ITIC computer voltage tolerance envelope (see Chapter 1) shows the relationship of overvoltage and undervoltage with respect to time and the three types of disturbances.

Type I transient overvoltage disturbances may have an indirect effect on operation. The effects, however, are unpredictable and it is difficult to identify them with actual disturbances within the Type I range.

Type II undervoltage disturbances, on the other hand, have a direct and predictable effect on electronic equipment. Computer manufacturers, therefore, must provide built-in sensors (usually on the DC side of the power-supply rectifiers) to power-down the equipment if voltage falls below the preset limits, and yet avoid nuisance tripping as a result of otherwise harmless low-magnitude transients, etc. In general, adequately large power supply filter capacitors can help reduce the effects of Type II undervoltage disturbances and momentary outages.

Acceptable input power requirements for critical loads vary with the type of equipment served and its manufacturer. Typical limits for computer-type equipment include the following:
- frequency within ±0.5 Hz;
  steady state voltage within ±8% of normal rated voltage;
- transient voltage conditions not exceeding +15% or -18% nominal, and return to normal voltage within 0.5 seconds;
- line-to-line voltage imbalance of a 3-phase system not differing by more than 2.5% from the average of the three voltages on each phase; and
- total harmonic voltage distortion not exceeding 5%.

When about to start a project, the above concerns must be addressed. Begin by obtaining from the serving utility a case history of disturbances that have occurred on their commercial AC power lines over the past five years. Ideally the data should be from the substation(s) that will serve the project under study.

A key aspect in understanding the scope of the project is open two-way communication between those designing a quality power installation and the client who will operate the facility. There must be a clear understanding of the client's needs, objective, priorities, and past experience in using critical power. Further, a clear understanding of the cost impact resulting from the failure of solid-state electronic equipment at the site will assist in evaluating the type of solution that will best meet the client's critical power needs.

It is always tempting to say that you can't go wrong by giving the client the "Cadillac" of the line. Actually, that is a fallacy that can lead to more problems than it solves.

DEFEATING NOISE AND OTHER PROBLEMS

Introduction of microelectronics into the workplace has increased efficiency dramatically. However, its use is a double-edged sword: along with the inherent benefits comes a specific challenge in the form of a requirement for clean and reliable power. Solid-state electronics is extremely susceptible to power disturbances.

As noted in Chapter 4, these disturbances may originate from sources on the utility or end-user side of the power distribution system. Severe weather conditions, lightning strikes, or accidents involving distribution lines typically characterize utility service problems. On the other hand, the end user's own power system can be a source of power disturbances. Equipment in a neighboring facility connected to the same power grid also is the possible source of electronic equipment operation problems.

Providing good quality power requires the use of many stratagems. The first level involves eliminating or minimizing outside sources of interference. The second requires the use of equipment and techniques to overcome built-in sources of problems that are inherent with the use of solid-state devices. For this purpose, the industry has developed an array of protective equipment such as power line conditioners, line voltage regulators, and UPSs to ensure dependable electronic equipment operation.

Choosing the best approach to solving the problems discussed in Chapter 4 requires a working knowledge of power disturbances coupled with equipment awareness. Specifically, how the equipment operates, what protective functions does it provide, and where it is to be applied.

COPING WITH NOISE

Noise cannot be "killed" but can be reduced to tolerable levels. It's, therefore, not meaningful to use words like "eliminate" or "stop" in their absolute sense. Each practical noise-reduction action should result in less noise where it is harmful, often by giving it an easier path elsewhere. Viable noise-reduction strategies consist of:

• raising the impedance to the noise in the harmful direction to reduce its amplitude;

• lowering the impedance in a benign direction to shunt it away; and

• breaking the link between the noise and the affected circuit.

For instance, noise currents will flow into a .01-ohm return path that has been provided and shunted away from a 1-ohm path through a solid-state electronic circuit by a ratio of 100:1. If only the 1-ohm path exists, all noise currents will be on it.

Ground loops happen most often in circuit returns. In **Fig. 1**, Loop 1 current (I_1) applies a voltage to Loop 2 because part of the circuit is shared. Where sharing of returns is not avoidable, such as by using a separate signal grounding system, a low-impedance return will keep the voltage down to manageable levels.

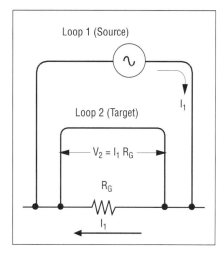

Fig. 5.1. Resistance of the section of the loop that is shared will determine how much voltage (V_2) is applied to target loop 2 by the current of the source loop 1. Lowering the resistance of the shared part of the loop or deducing the length of the shared part of the loop will reduce the interference.

Capacitive coupling. Capacitive pickup of noise can be a problem in high-impedance circuits. It often can be reduced by lowering the impedance of the target circuit. The reason that this effect is less dependent on frequency than might be expected from the relationship: $X_c = 1 \div (2\pi fC)$, is that stray circuit capacitance often creates a capacitive voltage divider.

Inductive coupling. Inductive pickup is more likely in low-impedance circuits. It is caused by the changing magnetic field created around a wire carrying alternating current and is characterized by the presence of two loops, one transmitting to the other as shown in **Fig. 5.2**. The larger the current is, the stronger the field will be. The area of each loop determines the effective potential for coupling, so flattening either the source loop or the target loop will reduce coupling, sometimes dramatically.

A magnetic shield does not block a magnetic field. Magnetic shielding will shunt the interfering magnetic field through itself and reduce the effect elsewhere. Doing all that is practical to reduce the coupling between the loops is still the best approach.

OTHER NOISE-REDUCTION APPROACHES

Other acceptable noise-reduction steps include any of the following approaches.

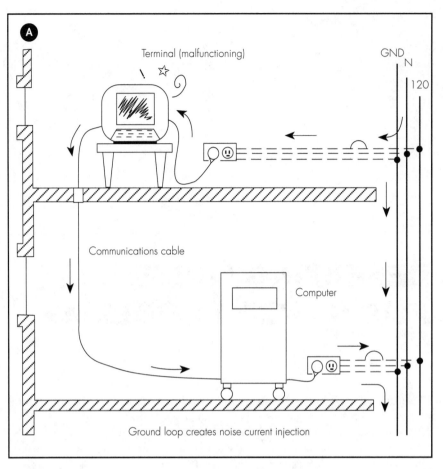

Fig. 5.3. (A) In high frequency noise is traveling in a wiring system and negatively affecting a PC terminal.

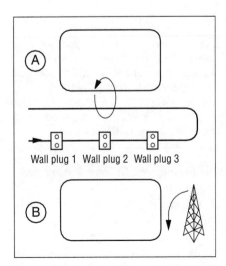

Fig. 5.2. A loop in an inductive field generated in a second loop is shown in (A). The inductive field strength is a function of current. A loop in an electromagnetic field (B) acts as a loop antenna.

Conductor sizing. Grounding conductor required sizes must be met or exceeded. Put in oversized ground conductors to the grounding electrode conductor and the power panels in the system. Any lowering of impedances in this location will help to shunt the noise coming through the transformer. Noise generated within the facility also is attenuated when it encounters a low output impedance.

Separation. Separate panels are preferable for power and for instrumentation/control use. Put motor drives, welders, and similar noise-producing equipment on their own power panels. Let the noise go to the low output impedance of the substation power transformer before it distributes back down to other circuits. This can easily lead to a 6- to 10-dB reduction.

Isolation. Isolation transformers should be used in place of standard transformers to step down voltage to the level needed in branch circuits. A standard transformer may have .002 microfarads of series capacitance from input to output, providing a highway for noise in both directions. A simple Faraday-shielded isolation transformer can reduce this series capacitance to 30 picofarads, nearly two orders of magnitude lower.

Filters. Shielded power-line filters should be used at cabinets and rack plugstrips to prevent noise on the power line from coming in and keep local noise from going out. The metal case of such a filter should connect with a low-inductance strap on the best signal-circuit return. This return should run, among other places, to the power panel ground because part of the filter's effectiveness in handling noise is to shunt it back toward its source.

DEFEATING NOISE ON WIRE PATHWAYS

Ambient noise in the wiring system (even at 1 to 2V) can interfere with the low-level signal voltages of the digital

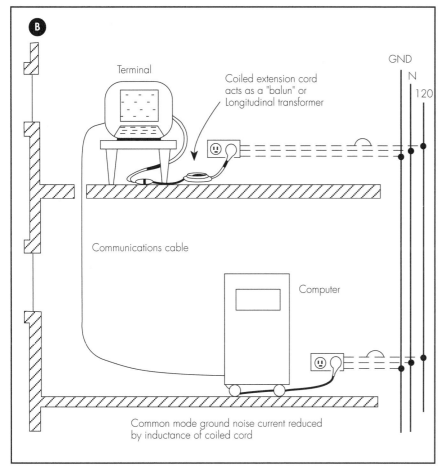

Fig. 5.3. (B) Air core coil is used to cancel or slow the effects of high frequency noise. It acts like a choke or balun.

equipment. The way the magnitude of interference is described is in terms of signal-to-noise ratio. As digital devices with lower logic voltages are introduced, this ratio is being reduced.

As was noted in Chapter 1, at one time the ratio was likely to be a very positive 30V signal to 1V noise (30:1). With this ratio, the possibility of interference with the signals was very low. Now, with this ratio is getting close to 1:1 and sometimes as low as 1:2, the digital signal system tries to run on the noise voltage level instead of the logic voltage. Finding no secure signal interface, the system reports to the main processor, "there is no signal," triggering a overall system shutdown.

One of the valuable tools used to cancel or slow the effects of high frequency noise traveling on the phase conductors is the coil or choke. Remember that an electrical coil resists a very fast voltage spike trying to force itself through the circuit. The term used for the coil's performance is "inductive kick." Actually, the coil develops a voltage that opposes the high-speed spike's travel, thus softening or reducing its effect on the circuit. The usual application of a choke coil is in series in a circuit where such a dangerously high spike might damage or destroy sensitive components. The device is similar to a balun (a small choke coil used in the telecommunications industry to protect signal circuits).

The same results are obtained from an air-core coil. This is done by simply coiling six to seven turns of an extension cord that may be in the power circuit of an electronic device experiencing circulating spikes coming down the ground wire and forming a ground loop.

A typical example is shown in **Fig. 5.3(A)** where the problem affects a CRT workstation. In **(B),** a temporary air-core coil is created to determine if the problem is really a high-speed voltage traveling down the grounding conductor. Once it is confirmed, if desired, it can be replaced by a longitudinal transformer (balun).

A balun may also be used to provide other functions, such as voltage stabilization or continuous energy supply to the load in question. In this function, the device fits into the category of power conditioning tools such as voltage regulators, line conditioners, motor generators, and UPS systems.

DEFEATING INDUCED NOISE

Electric and magnetic fields are associated with 60 Hz currents. Noise in one circuit couples itself onto circuits that are physically nearby. The same physical law that couples the primary and secondary windings of a transformer are at work here.

When current flows through circuit #1 in **Fig. 5.4**, a magnetic field is created that surrounds the wire. When the current is suddenly interrupted, the magnetic field will collapse. The moving magnetic field cutting through adjacent circuit #2 will induce an emf in

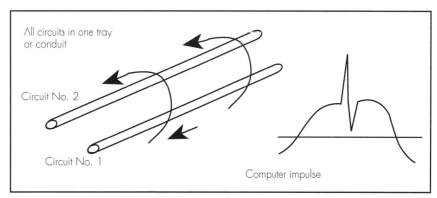

Fig. 5.4. Bursts of noise are coupled through the air from one wire to an adjacent one by means of an electric field, which is established by the flow of current in one wire (arrow along wire). It is coupled into the second wire by the electric field circling both wires.

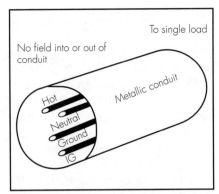

Fig. 5.5. A dedicated circuit has one hot wire, one neutral wire, and one or two ground wires, all serving one load and all enclosed in one metallic conduit.

that wire. The strength of the emf induced in the conductor depends upon the number of lines of force that cut through it per second; the greater the number, the higher the induced emf.

The induced emf in circuit #2 causes a current to flow. This unwanted induced spike appears as noise to the solid-state electronic equipment to which circuit #2 is connected. This induced noise has a more serious effect on power and signal wiring of sensitive systems. As discussed in Chapter 4, common-mode noise spikes heading into a coil generates an opposing force and the two forces in series produce a canceling or reducing effect.

If, however, these same two forces were in parallel, they would not oppose each other. Instead, they show themselves as separate spikes on the wire because there is no canceling.

Suppressing this noise. The most obvious answer is to impose some protective material, such as shielding, between the noise and the devices requiring protection. Shielding can take the form of running each circuit in a separate metallic conduit just as well as installing the conductor as part of a shielded cable.

Dedicated circuit. The recommended way to help cancel this induced noise effect is to use a dedicated circuit, as shown in **Fig. 5.5**. In this single-phase example, there is one hot wire, one neutral wire, and one or two ground wires, all serving one load and all enclosed in one metallic conduit. Two things have been accomplished with this installation. First, the individual circuits have been kept apart from each other. This is especially important if the circuits are serving dynamic loads that are subject to high-speed pulses or sharp changes in operation. Second, each circuit has been enclosed within its own shield, namely the metallic conduit. It will deflect and absorb the field energy.

Shielding can take the form of metallic conduit, shielded cable, or it can be shielded wire or even what is called "mu metal" to protect a workstation or terminal. Whatever the form used, the principle is the same: Block the effects of air-transported disturbances and redirect conducted disturbances.

One problem that arises when using shielded cable as the solution for eliminating ground loops that use the metallic shield as part of the circuit; grounding only one end of the shield. While this solution can solve that problem, it can cause even more serious ones from surges in AC power lines due to lightning.

Fig. 5.6 shows how a circuit having its shielding grounded at only one end can build up high voltages at the ungrounded end. In this example, a moderate lightning surge of 40,000A can lead to an 80kV potential between the shield and the remote terminal case.

The way of getting around this is to ground the ungrounded end of the shield through diodes as shown in **Fig.**

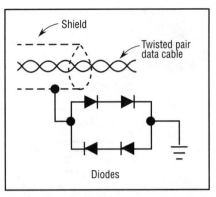

Fig. 5.7. By grounding the previously ungrounded shield end through diodes, the problem can be eliminated.

5.7. During normal conditions, forward voltage drop in the series-connected diodes blocks 60 Hz circulating current through the shield, preventing noise pickup by the data cable.

DEFEATING RADIATIVE NOISE

Electromagnetic (radio-wave) energy falls off as the square of the distance from the source. Transfer efficiency is low unless quarter-wave or longer conductors are involved. Shielding is tricky as can be seen by playing an AM radio near an idling auto engine. Signal strength is proportional to the RF current of the source, which is why resistor wires are used in auto ignition systems.

100 kHz to 1 MHz. In this part of the spectrum, circuit-return/common

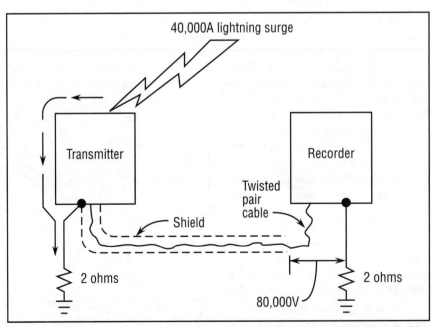

Fig. 5.6. A lightning strike near a field-mounted instrumentation transmitter allows buildup of dangerous voltage between the ungrounded shield end and recorder case.

Shielding can take the form of a large screen that covers all six surfaces of a room that's full of microprocessor-based digital equipment.

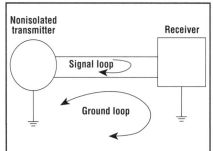

Fig. 5.8. Typical ground loop.

systems may begin to have spot resonances that reduce their ability to tie metallic objects together on larger systems over a few hundred feet apart; thus, methods of coupling that do not depend on low source-to-destination potential differences must be used increasingly. This means that shielded twisted-pairs, passive and active drivers, optical coupling, and even fiber-optic signal paths may be needed. Large-surface grounds between adjoining racks and pieces of electrical and electronic equipment are still effective. Antenna effects can be reduced by shifting alignment of the coupled circuits by subtracting from (or adding to) leads to avoid resonant effects; filtering the affected circuit so that the received energy is bypassed; or interposing a shield. This shield must not reradiate the energy or its effect may be to strengthen the interference. Also, the shield can sometimes be arranged to absorb energy by coupling to the source and dissipating the received energy in a resistive load.

1 MHz and up. At these frequencies, the only circuit return paths that can be counted on are those that can be seen. If the grounding common goes through bolted joints, especially if they are painted first, there may be no helpful connection at all. The best way to ensure a useful common is to install it intentionally. Locally, a 2-in. wide strip of copper 10 mils (.010 in.) thick or more will be helpful in many cases. Use liberal amounts of copper to tie chassis, racks, and cabinets together so that circuit returns are direct and predictable. Pay special attention to high-power systems at these frequencies (including arc sources), as they can radiate noise into other circuits.

AVOIDING GROUND LOOPS

A neutral-to-ground or neutral-to-signal-reference measurement is expected to be zero volts or close to it. If the measured voltage is much higher, it indicates the presence of a ground loop caused by grounding a device at two or more locations.

As shown in **Fig. 5.8**, a ground loop allows current to flow between the grounds by way of a process control loop, thereby adding or subtracting current to the process signal. The receiver is unable to differentiate between the wanted and unwanted signals and, thus, doesn't accurately reflect the measured conditions.

With so many connections within a facility referenced to ground, the likelihood of establishing more than one point is great. Thus, if an instrumentation system seems to be acting erratically, the chore of eliminating all unintended ground connections becomes overwhelming.

Remember, the types of problems that need to be dealt with are high-speed bursts of noise, or a high-frequency impulse usually seen between a neutral wire and the equipment case ground.

As with noise on phase conductors, a balun can be used here, but as a common-mode choke slowing down or softening the impact of the circulating noise. A shielded isolation transformer can also serve in the power circuit; this device will turn around the high-frequency burst that is traveling on the phase conductors and looking for ground potential. In effect, a shielded isolation transformer redirects common-mode noise back into the noncritical areas of a building's wiring. This device is a recommended modest-cost complement to any power circuit supplying digital power supplies.

When ground loops can't be eliminated, one solution is the installation of signal isolators shown in **Fig. 5.9**. They break the path between all grounds while allowing analog signals to continue through the loop. An isolator also can eliminate the electrical noise due to common-mode voltage.

IG-TYPE WIRING

One of the permitted methods for reducing electrical noise is the installation of insulated ground (IG) wiring.

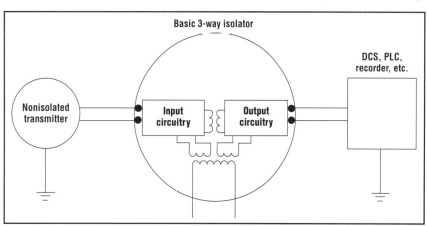

Fig. 5.9. Signal isolators break connections between all grounds.

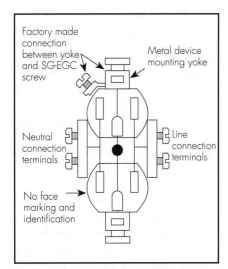

Fig. 5.10. Typical SG-style receptacle. Note factory-made connection between device yoke and greenwire screw.

Yet the truth is that about 50% of the time, it reduces electrical noise on the circuit to which it is applied; about 50% of the time it increases the noise; and the remaining amount of the time, there is no difference. It's important, therefore, that the technique be fully understood before it is applied.

Nobody seems to know what the notation IG means. This is no joke. Here, "I" will stand for "insulated," not "isolated." Thus, in this book IG stands for "insulated grounding." There is a real difference between the two terms. The difference can best be understood by examining the construction of receptacles.

SG receptacle. As shown in **Fig.**

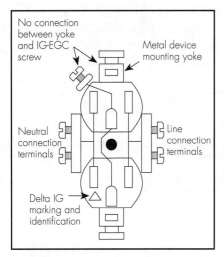

Fig. 5.11. Typical IG-style receptacle. Note lack of connection between device yoke and greenwire screw. Also not IG marking on receptacle face.

5.10, a solid grounding (SG) receptacle has both the equipment grounding pin and green-wire terminating screw made common to the metal device-mounting yoke. As a result, mounting the SG receptacle into a metal device box or item of equipment makes the receptacle's equipment grounding pin common to the device mounting box or equipment enclosure and, therefore, to whatever else these items are connected to on the building's installed grounding system.

IG receptacle. In an IG-style receptacle (**Fig. 5.11**), insulation is introduced between the equipment groundwire pin and the metal device-mounting yoke. This breaks the common connection described above for the SG design and, therefore, eliminates the solid-grounding connection it provides. It is now insulated, as opposed to solidly grounded at the receptacle end of the branch circuit.

An IG-style receptacle is identified by the presence of a permanently attached mark on the receptacle's face that is in the shape of a triangle, or delta. Any color may be used for the face of the receptacle. In the past, the IG-style receptacle was identified by its orange-colored face. Now an orange-faced receptacle may or may not be of the IG style. Aside from the marking, the manufacturer leaves out the connection between the equipment grounding pin/ wiring terminal and the metal device mounting yoke and puts the letters IG into the part number. These are always product safety listed by a nationally recognized testing laboratory (NRTL) and are required to be clearly identified as being of the IG style.

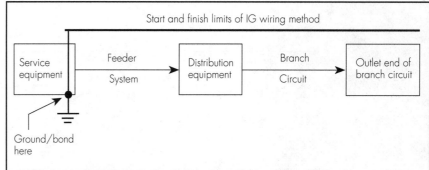

Fig. 5.12. These are the basic NEC-established premises wiring limits for the IG wiring method from the service equipment.

When installing an IG receptacle, an insulated green wire is connected to the IG receptacle's ground pin and at the other end, this wire is directly and solidly terminated to equipment ground at the service-entrance equipment or derived power source. Hence, the IG circuit is "insulated" from the downstream equipment to the receptacle; due to the connection at the downstream end, however, it has not been "isolated" from equipment ground.

IG WIRING METHODS

The NEC limits the range of the premises wiring system into which the IG wiring method may be installed.

Fig. 5.12. The IG wiring method is shown as being restricted to that portion of the premises wiring system extending from the solidly grounded AC system that is a part of the service equipment to the outlet end of the branch circuit. This arrangement requires that the voltage level at the service equipment be the same as that intended to appear at the outlet end of the branch circuit. Intervening electrical distribution equipment such as pull boxes, fused switches and disconnects, switchboards, panelboards, equipment cabinets or enclosures, and similar equipment are all part of the IG wiring system, and the green wire must not be connected to any of this equipment when it passes through them.

The outlet end of the branch circuit is the last part of the premises wiring system. It's the point at which the premises wiring system is finally interfaced to the branch-circuit-connected load equipment.

Fig. 5.13. A common variation of the

distribution system is where a separately derived AC system is interposed between the service equipment and the upstream terminating end of the IG wiring method. Here, a separately derived system, such as an isolation transformer, is used to establish another solidly grounded AC system. Here, the NEC requires that the output of the separately derived system (as opposed to the service equipment) becomes the furthest upstream point from which the IG wiring method is run downstream to the electronic load.

This method permits the use of a higher level distribution voltage from the service equipment, since the isolation transformer can be of the stepdown variety. Also, the isolation transformer may be positioned very close to the load equipment served by the IG wiring method; this has significant advantages in many cases.

Fig. 5.14. When the AC system supplying the IG wiring method consists of more than one AC source, such as when an uninterruptible power supply (UPS) is used along with its associated transfer/bypass switching, things get much more complex from an equipment grounding standpoint. Arrangements like this involve AC systems that are classed as being of the "solidly interconnected" AC system type, as opposed to being of the separately derived AC system type.

Note that in this arrangement there is no AC system grounding shown at either the UPS or at its associated transfer/bypass switch. Instead, all the AC system grounding occurs at the service equipment. This happens since the neutral (X0) conductor from the UPS output is routed upstream to the service equipment, where it's solidly interconnected to the neutral bus bar. This con-

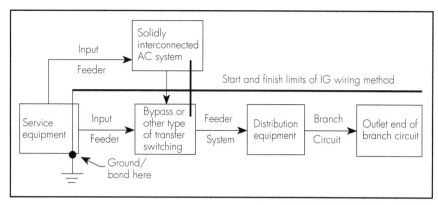

Fig. 5.14. In a solidly interconnected AC system arrangement (common with UPS systems), the IG premises wiring limit extends upstream to the service equipment.

ductor is routed within the UPS input feeder, along with the line conductors supplying the rectifier-charger. This is what makes the two AC systems (service equipment and UPS inverter) solidly interconnected.

These solidly interconnected arrangements utilizing IG wiring techniques usually involve a direct connection of the grounding conductor to the equipment (rather than through a receptacle). Here, the green wire is directly connected to the grounding terminal of the electronic equipment, which in turn also grounds the enclosure. The other end of the green wire is connected to the ground bus at the service equipment or separately derived source.

In many cases, however, the direct connected IG wiring involves running the power feed conductors and the green wire from the power source to the electronic equipment enclosure via a metallic conduit or raceway. If the raceway is rigid metal conduit, IMC, or EMT (suitable for use as an equipment grounding conductor path) the IG feature would be defeated if the raceway is connected to the enclosure. This would, in effect, tie the two grounding systems together. Thus, the raceway must be connected to the enclosure by an insulating bushing.

This means of IG connection is relatively new to the NEC and is covered in the Exception in Sec. 250-75. Basically, its intent is to give the same opportunity of using IG-style wiring on direct-connected electronic load equipment as cord, plug, and receptacle-interfaced versions have had.

OTHER IG WIRING TIPS

Permitted NEC use. There has been a lot of confusion over where the IG-style connection can and cannot be used in an electrical system installed under the requirements of the NEC. Note that there is only one location that is acceptable: at the outlet end of the branch circuit used to connect to electronic load equipment. Any attempt to place an IG-style connection (receptacle or direct connection) onto any other point on the wiring system, or to serve non-electronic equipment with it, is an NEC violation of either Sec. 250-74, Exception No. 4 for receptacles, or Sec. 250-75, Exception for direct connections.

Effect of impedance. Loads that are affected by common-mode currents and voltages propagated over the AC power branch circuit fall into high- and low-impedance categories. High-impedance loads are susceptible to electrical noise in the form of voltage. Low-impedance loads are susceptible to electrical noise current.

The IG wiring method changes the lower impedance of the typical SG (solid grounding) branch circuit into a much higher impedance. Hence, it attenuates noise current, but permits an

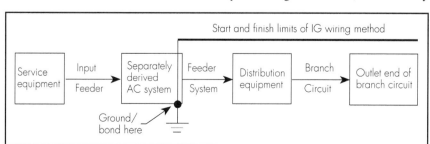

Fig. 5.13. These are the basic NEC-established premises wiring limits for the IG wiring method used with a separately derived AC system.

47

increase in noise voltage on the path.

Looking into the load-end of an SG branch circuit, a relatively low impedance is seen that permits a fairly heavy noise current to be propagated along the path, but with minimum voltage drop. Conversely, looking into the load-end of an IG circuit, the reverse condition is seen: maximum voltage drop and minimized current flow.

General rule. To improve electronic load immunity or to reduce susceptibility to common-mode electrical noise, use inverse connections (connect a low-impedance load to a high common-mode impedance branch circuit, and a high-impedance load to a low common-mode impedance branch circuit).

Effect of length. The length of the IG circuit from the solidly grounded AC source to its outlet end affects the impedance of its path from one end to the other. This, in turn, affects the amount of electrical noise current (as opposed to electrical noise voltage) the IG path can propagate along its route. Whatever effect the IG wiring method is going to have with a given load will be proportional to the length of the IG circuit. A similar effect exists on the SG circuit, but in inverse fashion.

From this, it's apparent that placing an IG receptacle 6 in. away from the branch circuit panelboard, and having the panelboard a few feet from its serving solidly grounded transformer that forms the separately derived AC system, is not going to have much effect at all. There simply isn't enough IG length to the circuit to develop any useful common-mode impedance on the equipment grounding conductor path for the arrangement to have an effect. However, if the IG branch circuit is 100 ft or 200 ft long, then the desired effects would be appropriately produced.

In general, there's no useful gain in installing IG receptacles onto the same case/enclosure that contains a solidly grounded power conditioning device such as a transformer used with a UPS, a voltage regulator, or similar item of self-contained power conditioning equipment.

Potential differences. Another important fact to consider is that if the site has a lot of potential difference between points of grounding (as over large area buildings), an IG wiring method will ensure that the AC power source and the electronic load get conductively connected across that potential. The result is increased common-mode current trying to flow on the IG path and a much greater common-mode voltage being developed across it. This can cause more interference with the victim load, if it has a communication or signaling circuit attached to it and is ground-referenced at some further distance away.

Effects of lightning. IG wiring methods also bring with them increased susceptibility to lightning-related problems. When lightning strikes the face of the earth, there's a step-potential (kV/meter) created across the soil. Similarly, this kind of potential difference can be created by lightning currents vertically, horizontally, and on any diagonal in a facility of given size, particularly multistoried and large area buildings. Hence, there can be a very great potential difference, for example, between a service entry main grounding point and any remote point in the building when a lightning strike occurs.

This means that if the victim load equipment is located a long distance from the service entry, its local grounding conditions will be at a significantly different potential from what the conditions at the service equipment will be. If this path and its potential difference is then bridged by the connection of an IG equipment grounding conductor, there will be a traveling voltage and current wave front barreling down the IG conductor in an attempt to permit current flow and to equalize potential between the two points. This is called a lightning surge current.

Assuming the service entry is the originating point for the surge current and the victim electronic load is the terminating point for it, if there are data, signaling, or communications cables attached to the load equipment, the lightning surge current will try to flow on these circuits in a continuing attempt to find "ground." Typically, this is quite damaging to these circuits and, since the surge current is in the common-mode, no amount of line-to-line and line-to-ground surge protection equipment installed on the AC power circuit will protect the load and its attached circuits.

Also, if there is grounded metal near the victim load, and the surge current has sufficient potential driving it (due to impedance mismatch and related voltage reflections at the load-ground interface point), there can be a lightning sideflash of up to 6 ft horizontally through the air between the energized victim load and whatever grounded materials or other equipment is nearby.

If an SG circuit is used in the above case instead of the IG style, the victim load will be largely protected, since it will have become ground-referenced to its local ground conditions and will be generally unable to develop a side-flash potential problem. The same holds true for an IG circuit that is physically short and its AC system is not remotely located, but is instead a dry-type transformer (or similar source) installed as a solidly grounded and separately derived AC system in the same location as the load being served on the IG path.

IG vs SG. There are no clear winners and losers between the IG and SG wiring methods. Therefore, no blanket recommendation or statement can be made in relation to the SG or IG methods as regards to predictable benefits that either can provide. The results are related to the site's conditions, the length and routing of the circuit, and the high- and low-impedance nature of the victim loads themselves. Also, the effects of the SG wiring method versus the IG wiring method on electrical noise are only useful with common-mode noise problems, not transverse-mode ones.

POWER CONDITIONING

When the simple strategies mentioned thusfar are not sufficient to meet the requirements of the solid-state devices for noise-free power, power conditioning equipment must be employed.

Power conditioning is often thought of as the act of taking "dirty" incoming power, passing it throughout a piece of equipment, and ending with "clean" power. Reality is that the result very often is that the sine wave containing a moderate amount of noise is conditioned and the consequence on the output side is a waveshape that cannot be tolerated by the solid-state electronic load it was intended for.

The reason for this seeming contradiction between purpose and result is that the input power, power conditioning equipment, and load are dynamic and interact with each other. That is why it is necessary to first understand the nature of the load and the proposed power conditioning equipment in order to determine their compatibility.

The topic of power conditioning is very complex and the equipment that can be used for this purpose is varied. For that reason, power conditioning is discussed fully in Chapter 8 of this book.

INCREASED SERVICE RELIABILITY

Power reliability, as was discussed earlier, is a major factor in assuring that solid-state electronic devices operate properly. Although UPSs and other power conditioning devices can be used to overcome power sags or outages, this solution is generally limited to short periods of time. When longer outages are involved, alternate power sources such as engine generators must be brought on line.

Electric utility companies also are aware of the requirement for reliable power for their customers who have large mainframe or process-control computer installations, or large quantities of PCs or other solid-state electronic equipment in their installations. They, therefore, have made significant efforts to increase the reliability of their electric services. Here are some of the equipment that has been developed for this purpose.

Low-voltage subcycle transfer switches. Power failure is common to all electrical equipment and was addressed by using switches to transfer from a "preferred" to an "alternate" source. The transfer switches used were typically electromechanical. They operated on the principal that a reduction in voltage below the dropout value of a relay would cause a relay to deenergize, disconnecting the preferred contactor and energizing the alternate contactor.

Since the two sources might be out of phase, means had to be provided to insure the preferred source was completely disconnected before the alternate source was connected. This was necessary in order to prevent an interphase short circuit, and was accomplished by introducing a time delay after dropout before transferring to the other source.

For supplying solid-state equipment, this approach is not usable for the following reasons:
• power interruption is too long;
• arcing during transfer is a source of intense radiated and conducted noise; and
• large current transients are produced by uncontrolled transfer switching.

As an offshoot of the UPS, a subcycle automatic transfer switch (ATS) for stand-alone applications at 600V and below was developed. The device is capable of a 4-ms transfer, with a zero crossing at the midpoint of the transfer, thus avoiding any reverse current flow when the transfer takes

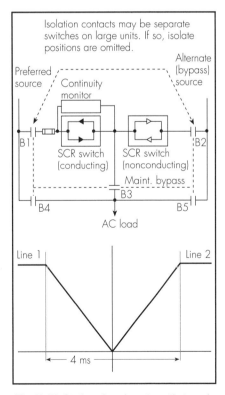

Fig. 5.15. *Static subcycle automatic transfer switch allows 4-ms switching between two energized power lines, with a zero crossing at the midpoint of the transfer, thus avoiding any reverse current flow when the transfer takes place.*

place. This application has found favor with users having two independent power sources at 600V and below; they now could count on these energy sources to back up each other. (A typical "stand-alone" configuration is shown in **Fig. 5.15**.)

The beauty of this concept is the potential elimination of a battery-support system, along with the associated high cost of rectifier/inverter assemblies. By using a solid-state subcycle transfer switch, 75% to 80% of installed UPS cost can be saved. In fact, many dual-feed locations such as hospitals, computer service centers, banks, and insurance operations have made use of this technology, reducing their expense for continuous service to as little as a quarter or a fifth of the cost of an appropriately sized UPS.

Certain prerequisites, however, must be considered to ensure the above application is a valid one. First, two power sources must available. Second, no high shock torque must be applied to downstream motors during the operation of these solid-state subcycle devices.

These transfer switches are available in sizes up through 4000A and capable of handling up to 3000kVA of load. Equipment costs for this low-voltage technology are in the range of $500 to $900 per kVA. (Smaller sizes are usually at a higher cost per unit.)

Space considerations are just as important as equipment costs, however, especially when tenant occupancy costs in major buildings run from $30 to $75 per sq ft. Here again, the solid-state, subcycle transfer switch provides a cost benefit: Along with a much lower cost of maintenance, it takes up considerably less room than a UPS and its peripheral support equipment. By eliminating the needed space for a large battery system, equipment cabinets, etc., real savings are achieved. In several static switch installations, the recovery of space cost (through additional rental revenues) has made construction and material financial arrangements even more attractive.

MV subcycle transfer switches. Further development of power level silicon technology extends this capability of high-speed transfer between two

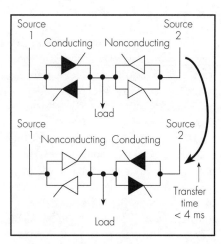

Fig. 5.16. Silicon transfer devices operate as gated thyristors. The "gate: is the switch's control device, which issues the signal to the silicon to conduct or turn off. These functions are done in microseconds, so the speed of transfer does not interfere with sensitive loads.

power sources to the following MV distribution voltage ranges: 2.4kV, 4.8kV, 15 kV, 27kV, and 38kV.

In 1994, with increasing interest from primary service customers, one utility began a process that led to the installation and operation of the first 15kV, 600A, 14MVA, 4-ms switch. Resultant testing and field verification have shown consistent operation and performance characteristics that are all within design parameters. This installation has led a number of utilities to order these MV devices for their systems and specific customer installations.

Fig. 5.16, shows a typical transfer from one incoming line to a second one. There is very little difference between the low-voltage and MV versions. First, both operate as gated thyristors. The "gate" here is the switch's control device, which issues the signal to the silicon to "conduct" or to "turn off." These functions are done in microseconds, so that the speed of transfer will never interfere with sensitive loads. Second, both versions have no moving parts, so that repeated transfers will not cause equipment deterioration. Finally, the use of an established technique (the gated thyristor) adds higher speed sensing and control advances.

One innovation recently applied to subcycle transfer switching is the control of the gating signals by fiberoptics rather than shielded copper wire. The result is a lighter and smaller equipment package and more secure signal transmission.

The control system for the MV version includes digital signal processing (DSP), which increases the speed and quantity of information used to perform the transfer functions. For example, when the MV controller senses an irregularity on Line 1, it collects information from 192 points per electrical cycle. It then uses 2.5 ms to store more than 30 points of information in making the decision to transfer to Line 2. One of the deciding factors in a transfer is the presence of a downstream fault; this condition needs to be cleared by circuit protective devices. This is evaluated to prevent initiation of transfer to the second line, which would only feed the faulted portion.

Another innovation is the use of the MV version in a metalclad or metal-enclosed lineup of switchgear. A complete lineup, including electromechanical back-up transfer switches, is really a replica of the UPS function. In **Fig. 5.17**, a conventional electromechanical MV transfer switch (with motor operators designated by the letter "M") is used in a bypass arrangement. Here, the silicon devices are the primary transfer devices; however, the static and electromechanical control systems exchange information with each other to ensure automatic operation. If the silicon switch needs to be taken out of service, the bypass device (electromechanical transfer switch) is signaled to close in parallel before the silicon device is opened. Continuity is just like a UPS with its bypass circuit.

Static series voltage regulator. Although the term "power quality" is not often associated with higher voltage levels, the concept is just as important for medium- and high-voltage systems as it is for 600V systems. In fact, electric utilities and their research consortiums have devoted time and funding to improve service technology at the primary voltage level. The end result has become known as "custom power."

A recent study, the Distribution Power Quality Monitoring Project, conducted by the Electric Power Research Institute (EPRI) with the cooperation of 24 utilities, involved the monitoring of 277 sites. It produced information on voltage sags and swells, transient disturbances, momentary disruptions, and harmonic resonance conditions. This was discussed previously in Chapter 4, and a portion of the data was shown in Fig. 4.13.

A large percentage of sags and momentary outages (the data found at time durations of 1 sec or less) were observed. These short-term power system

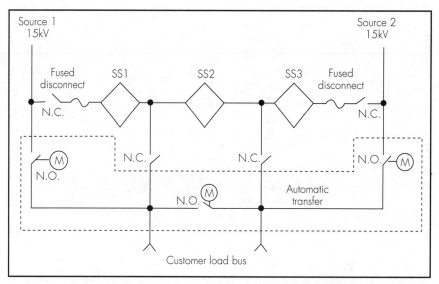

Fig. 5.17. Conventional electromechanical MV transfer switches (M) are used in a bypass arrangement, with the silicon devices being the primary transfer devices. The static and electromechanical control systems exchange information with each other to ensure automatic operation. If the silicon switch needs to be taken out of service, the electromechanical bypass device is signaled to close in parallel before the silicon device is opened.

disruptions weren't as troublesome when analog devices were the norm for end users. Now, almost every major voltage sag and momentary breaker operation causes loss of system control or a complete production shutdown.

Faced with increased end user equipment sensitivity, utilities redoubled their efforts to develop technologies dealing with sags and momentary outages. One result of these efforts is the static series voltage regulator (SSVR).

When a phase-to-ground fault occurs on an MV power system, feeders near the faulted area experience phase voltage distortions. If these distortions last for several cycles (which seems to happen frequently, as the EPRI data indicates), solid-state electronic loads are negatively affected. With an SSVR, if phase voltages drop below a preset threshold, it will boost the voltage to support continuity of the correct voltage to a critical load area. And, with proper sizing, an SSVR will support other voltage drops, such as those resulting from momentary breaker operation. This is not protection from a complete loss of power (as with a UPS) but rather protection against short duration problems.

There's still a danger in the application of an SSVR that an upstream disturbance on the utility system might affect both independent power sources at the same instant of time. After all,

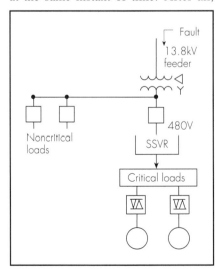

Fig. 5.18. One result of utility development efforts is the static series voltage regulator (SSVR). If phase voltages drop below a preset threshold, it will boost the voltage to assure correct voltage level to a critical load area.

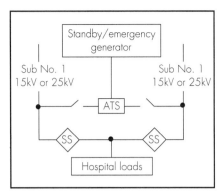

Fig. 5.19. An example of a multiple/redundant service includes a new high-speed transfer device between two existing substations to keep the facility on line in the normal manner. If an incoming line is interrupted, this transfer switch transfers to the alternate feeder in less than 4 ms, thus keeping the load on line. The existing engine-generator set is brought on line and made available to subtitute for the lost incoming line.

how far back into the utility grid can "independency" be maintained? Surely, these lines must come together somewhere.

This possibility already has been anticipated. The result is a combination of the "boost" concept (shown in **Fig. 5.18**) and the high-speed subcycle transfer switch discussed previously. This works very well in providing fast transfer between lines while maintaining correct voltage should a common disturbance on the upstream circuits (in this case, a single exposed transmission line) occur.

Multiple/redundant services. Many facilities, such as hospitals and health care centers, have existing dual feed services. Another alternative for reliable power is a scheme called multiple/redundant services. As shown in **Fig. 5.19**, a solid-state subcycle transfer device is installed between two existing substations to keep the facility on-line in the normal manner. If an incoming line is interrupted, this transfer switch transfers to the alternate feeder in less than 4 milliseconds, thus keeping the load on-line. The existing engine-generator set is brought on-line (with a 1-sec to 2-sec time delay to avoid false starting), synchronized, and made available to substitute for the lost incoming line. The net result: A restoration of the

An SSD, including its cryostat can be installed in a semitrailer allowing it to be moved from one location to another, as needed.

two-source capability within several seconds. The possibility of a cascading failure of both lines is avoided by bringing the third source (the gen-set) on-line after the loss of the first source.

SUPERCONDUCTIVITY

Until recently, there has been no satisfactory remedy to eliminate short but costly disturbances affecting large industrial loads. A new application of superconductor technology, the Superconducting Storage Device (SSD) offers up to 5 MW of power for critical loads.

Superconducting materials are characterized by a lack of electrical resistance, and can carry large currents without losses. To achieve this performance, low temperatures are required. The SSD's coil is submerged in liquid helium contained in a vacuum-insulated cryostat.

Fig. 5.20 shows one of the many configurations in which the SSD can be connected into a system. In this case, it is intended to carry multiple and varied loads during voltage disturbances on the utility power system.

The SSD stores energy within a magnet that is created by the flow of DC current. In the standby mode, current

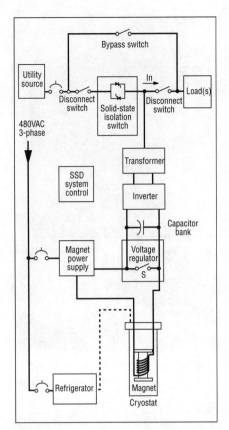

Fig. 5.20. A shunt-connected superconducting storage device.

circulates through the normally-closed switch (S). The power supply provides a small trickle charge to replace the power lost in the non-superconducting part of the circuit. When the voltage across the capacitor bank drops during a sag or outage, switch (S) opens and current from the magnet flows (in this case) to the inverter and thus on to the load. The system controller opens the solid-state isolation switch within 2ms to prevent backfeed.

When the utility power source returns to normal, the load is resynchronized within nine cycles before the isolation switch closes.

GROUNDING FOR SENSITIVE (SOLID-STATE) EQUIPMENT

Shutdown of sensitive electronic equipment is often attributed to the lack of quality power. Too often the immediate response to the problem is to think in terms of hardware that will condition the power directed to the equipment. This approach sometimes leads to the solution of the problem, but many times it does not. Other more basic problems may exist that defy the ability of any power conditioning equipment to correct.

First and foremost, it is necessary to analyze the power distribution system within the facility to see if an adequate platform for supporting such a load exists. The principal culprit is often found to be noise transmission via the grounding system in circuits supporting information technology equipment including, but not limited to, computers, peripherals, instrumentation, and communications equipment. The fundamentals of what constitutes a good grounding system for this solid-state electronic equipment will be discussed in this chapter.

IMPORTANCE OF GROUNDING

The job of making 5V microprocessor circuitry and other sensitive electronic circuitry reliable is daunting. A consistent overall approach is required, yet instruction and advice from different equipment manufacturers often is contradictory or vague, sometimes compromising safety or operating integrity, or both, even unwittingly.

The NE Code is concerned with electrical safety and does not address the problems of noise control or signal reliability. Building codes mandate NEC-based installation practice for safety. To operate microprocessors and electrical power devices together, however, it is necessary to follow rational design practices to reduce noise. These two requirements are not necessarily the same. Does this mean that NEC practices must be circumvented to avoid noise? If the end result is to make a system unsafe, then the answer is no. Even if compromises are helpful (and that is often not provable), they are not worth the risk of hurting someone. Safety is first and foremost!

The NE Code is concerned primarily

Control rooms *are the nerve centers of large installations containing computers, instrumentation, communication systems, and other items employing sensitive electronic equipment. Grounding techniques applied must assure that ground loop currents are not permitted to circulate, while also assuring personnel safety.*

with 60-Hz power circuits, with occasional consideration given to DC and 415-Hz power systems. Noise exists up to many megahertz. Thus, the best approach for designing a safe and noise-free system is to specify all the required NEC safety grounds; then, without circumventing this foundation, follow procedures to cope with the noise.

Installing separate ground systems is a dangerous practice. Separate power grounds can quickly become illegal and dangerous, but safe and effective signal grounding can be accomplished easily by using methodology contained within the NE Code!

A single-point ground is not an arbitrary point. Best results will be obtained when this single point creates the shortest path for the highest system currents. It could be a point on a process line. An "earth ground" connection (in addition to any required safety ground) is optional in this case. The essence of this technique is not the ground, but to have one place to connect all low-impedance commons.

EFFECTIVE COMPUTER GROUNDING

All sensitive electronic equipment must be properly grounded to assure both safety and proper operation. Large computer installations, such as those for data processing or process control, need special attention because they consist of many separate pieces of equipment that are interconnected and the units are often located within a dedicated room. The method of properly achieving good grounding this type of equipment, thus, can be complex.

To prevent electrical noise from affecting computers and other equipment containing solid-state devices, two entirely different and separate grounding systems are required. First, the power distribution system must be grounded in accordance with NE Code requirements for safety; and second, the computer equipment and enclosures must be connected to a signal reference grid (SRG) system for high frequencies. This high-frequency protection is in no way related to system grounding! In fact, the high-frequency reference will work whether or not it is grounded to the system ground. However, since it consists of exposed, metallic noncurrent-carrying parts that could accidentally become energized, it must be grounded, as required by NEC Sec. 250-42.

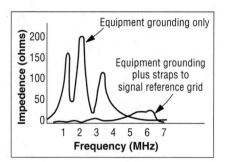

Fig. 6.1. Effect of a reference grid on grounding impedance.

Once it is recognized that proper computer grounding requires both a 60-Hz and a high-frequency grounding system, and that electrical safety requires that the two systems be bonded together, each system can be examined separately and then consideration given to how they are to be bonded together so they are compatible. **Fig. 6.1** shows the effect of an SRG on grounding impedance.

POWER SYSTEM GROUNDING

All equipment grounding essentially requires that exposed noncurrent-carrying metal parts be connected by equipment grounding conductors to a grounding electrode at the electric service equipment, or at the power source if power is obtained from a separately derived system. The first step in providing an adequate grounding system for solid-state equipment is to decide on the type of grounding electrode to be used for both power-system wiring and for the SRG.

The NE Code considers the following items to be suitable as grounding electrodes:

• metal underground water pipe in direct contact with the earth for 10 ft or more;
• metal frame of a building that is effectively grounded;
• concrete-encased electrodes within and near the bottom of a concrete foundation or footing in direct contact with the earth;
• ground ring encircling the building or structure in direct contact with the earth at a depth of not less than 2½ ft below grade level;
• rod and pipes not less than 8 ft in length; and
• plates with not less than 2 sq ft of surface exposed to exterior soil.

When considering the type of grounding electrode system to install in a facility containing an SRG, the most appropriate of the permitted types is the ground ring. In order to distribute and equalize any current flow into the earth, the grounding electrode system should completely encircle the building, be bare, and be no smaller than No. 2 AWG copper. If the structure containing the SRG includes steel columns, then these columns must be bonded to the ground ring. This is necessary because grounded structural steel is a qualified electrode and the Code requires all elements of the grounding electrode system to be bonded together.

A power distribution unit (PDU) is an item often used in computer/control rooms to provide power to solid-state electronic equipment. They come in a great number of variations, but generally include a transformer and a panelboard combined within the unit. **Fig. 6.2** shows a typical PDU. Care must be taken to provide the proper grounding for the unit.

The branch circuit supplying this equipment must meet all Code rules for equipment grounding. The equipment grounding conductor must run with or enclose the circuit conductors, per Sec. 250-91(b). For reliability and to reduce potential differences, a separate equipment grounding conductor of equal size to the line conductors should be run in parallel with the raceway, bonded to it at both ends.

Do not try to ground one of these units through anything other than the normal equipment grounding return path that runs with the supply conductors. Any fault return current in an AC system that is forced to flow through a separate path than the supply conductors will experience a much higher impedance, with the likely result that the fault won't clear as it should. In addition, such "isolated" or "special" grounding systems usually are counterproductive, inserting additional common-mode disturbances into the distribution system.

Sec. 645-15 frequently applies to this type of equipment. The second sentence reads as follows: "Power systems derived within listed...equipment that supply...systems through receptacles or cable assemblies supplied as part of this equipment shall not be considered separately derived for the purpose of applying Section 250-5(d)." It is worth looking carefully at what this sentence does not say.

It does not say that PDUs are not separately derived systems. They are, and as such, they are subject to the rules of the Code, per the definition of premises wiring in Article 100. The sentence merely says that a particular requirement in the Code [Sec. 250-5(d)] will not apply. The rule does not apply to general branch circuit wiring, only to cables and receptacles tested and evaluated as part of the unit.

Finally, the requirements of Sec. 250-51 for an effective ground path are not waived. The XO terminal will be bonded to the equipment ground, unless the equipment is double insulated, or unless a qualified testing laboratory has implicitly evaluated the ability of a fault to be cleared by the overcurrent device protecting the power supply. The first FPN indicates that product standards are safe.

PDUs must be grounded to the ground-plane for the raised floor. A grounding electrode conductor running without splice to a preferred electrode won't be required by the Code. However, the equipment grounding system that originates from this equipment must be at an equal potential to that of the other equipment in the room, or all the effort to properly bond the area will be for nothing. Make the connection to the grid in addition to the normal grounding return path.

Use of insulated ground receptacles needs to be carefully thought out on the load side of the power center. Although they will help make the PDU a single grounding point for the system, that branch circuit isolation may or may not be useful when the entire room has been bonded as discussed. The safest course is to consider IG receptacles on a case-by-case basis. The subject of IG receptacles is covered in Chapter 5 of this book.

HIGH-FREQUENCY GROUNDING

Computers and other solid-state devices input, process, and output data at tremendous rates, which are constantly increasing. Typical computer clocks that control data transfer operate at about 10 to 30 MHz. Some high-speed computers operate at several thousand MHz. These are radio frequencies (RF), and solid-state devices are sensitive to these frequencies. The wiring can act as a receiving antenna and can respond to external RF signals, which may cause false data to be processed.

Computers process data in square-wave pulses with rapid rise times, and these pulses plus the square-wave harmonics have frequencies up to 100 MHz or more. They may also generate RF signals that can radiate over the wiring, which acts as a transmitting antenna. In fact, the FCC has established maximum permissible signal radiation levels from computers. The RF signals that can radiate from or get into computers are known as electromagnetic interference (EMI).

Microprocessors operate at low voltages—usually 5 to 12V. Therefore, it is critical that unintentional voltage differences between various peripherals in a computer/control room be extremely small. At 60 Hz, this is relatively simple—just tie them together with a low-resistance, low-impedance grounding conductor so the enclosures and grounded metal are all at the same (ground) reference potential.

At radio frequencies the solution is not so simple. Low-impedance grounds are not easy to obtain because the inductive reactance of a conductor is proportional to the frequency. At 30 MHz, a length of conductor has an inductive reactance 500,000 times the reactance at 60 Hz. In addition, there is stray inductance and capacitance from conductor to conductor, or conductor to grounded metal, and resonance effects at high frequencies. This makes it very difficult to get a conductor of any appreciable length to have the same voltage at both its ends. If there is a difference in potential between the ends of a grounding conductor connecting two pieces of data processing equipment, it is possible for data errors to occur.

A high-frequency impulse in a conductor travels along the conductor at about 85% of the speed of light until it reaches the end of the conductor. Here it is reflected back along the conductor, becoming a traveling wave. At some frequencies, the reflected waves meet and reinforce oncoming new waves, creating resonances and standing waves. At or near these resonant fre-

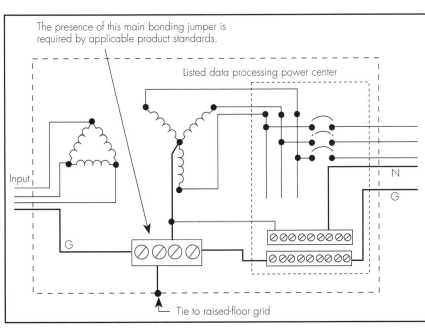

Fig. 6.2. Schematic diagram of a typical PDU for information technology equipment.

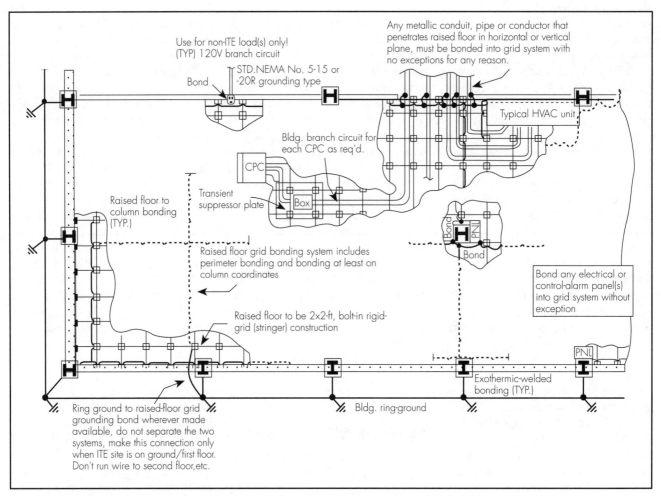

Fig. 6.3. Grounding and bonding of a raised computer floor and peripheral equipment.

quencies the conductor appears to have an extremely high impedance and does not provide an effective voltage equalizing means between two pieces of equipment. In addition, the conductor can act as an antenna at or near these resonant frequencies, radiating energy that can interfere with other equipment or receiving stray signals from other sources and presenting a false voltage signal to the computer equipment.

These effects are completely unpredictable, because the interfering signals are not steady, and because microprocessor-based equipment sensitivity varies. Digital circuits operate on binary signals, ON or OFF, (0 or 1). The circuits are most sensitive at the moment they are switching from one state to the other. If an impulse occurs at that moment, a false "bit" of data can occur in the system. At other times, the same impulse may not affect the processing. These errors can be extremely difficult to identify.

This points out the need to interconnect all cabinets of the system in such a way that they are at the same potential for all frequencies, from 60 Hz or less to very high radio frequencies. One of the best ways of accomplishing this is with an SRG.

ESTABLISHING AN SRG

If conductors are connected in a mesh or grid to form a multitude of low-impedance loops in parallel, there will be little voltage difference between any two points in the grid at all frequencies from 60 Hz up to a frequency where the length of one side of the square represents about 1/10 wavelength.

If such a grid is installed in a computer room and the enclosure or frame of each piece of equipment is connected to it by a short conductor, there should be no noise problems resulting from differences in ground potential between

A properly configured raised floor can be used as a reference grounding grid.

the frames of any two pieces of equipment. The preferred materials are braided copper, followed by copper strap. Round conductor is the least desirable, because it provides the least-effective results with high-frequency signals.

The best way to begin is to create, as perfectly as practicable, an equipotential grounding structure for the computers, peripherals, and other solid-state devices. The described system can usually be retrofitted into an existing installation, and will work with either 60 (or 50) Hz or 415 Hz systems.

Fig. 6.3 is an illustration of an arrangement where equipment won't notice the disturbance when it gets there. This is accomplished by minimizing potential differences. If there is almost no potential difference, then there will be almost no current. To the extent there is no current from outside influences, then there will be no effect on the logic circuits in the equipment.

This approach goes beyond NEC requirements. The mission of the NEC, per Sec. 90-1(b) is not to ensure that an electrical system will work, but to ensure that when wired, the system won't cause a fire or electrocute someone. The method presented here should allow all information technology systems to work properly, and to ensure that its operation will be safe.

Using the raised floor. Standard 2 × 2-ft raised floor will work well, but it must be the type that bolts rigidly together at each intersecting point. If a rigid-grid floor system can't be used, then a substitute will have to be created for grounding purposes. This option will be discussed later.

A 2-ft grid spacing assures that the grid won't become resonant with a high frequency signal, turning it into an antenna for unwanted electronic noise. This will happen if the conductor is a significant fraction of the wavelength of the signal received, and this must be avoided. The wavelength is the speed of light (32,108 m/sec) divided by the frequency. At 30 MHz, that turns out to be about 10 m, or about 33 ft. Even if the room is that long or longer, the 2-ft bonding intervals create a collection of shorted-turn cells that form a two-dimensional network of impedances for all current flowing in the grid. Due to their length, the shorted-turn cells are not normally self-resonant at any frequency of concern for commercial grade solid-state devices.

The purpose of the equipotential grid in the room is to make sure everything rises and falls together in the event of a power system disturbance, from whatever source. Begin by drawing an electrical fence around the area using No. 2 AWG (or larger) copper, which is the same size as the minimum for the ground ring electrode. This conductor must encircle the entire perimeter of the raised floor. It must be bonded to each pedestal it passes, and to each building column wherever possible. This conductor can be laid on the structural floor or can be suspended by appropriate fittings from the floor pedestals.

If the room is large with internal columns, then pick up those internal columns by using them as points of reference to subdivide the area. Go column to column from the perimeter bonding conductor with additional runs of No. 2. As in the case of the perimeter bonding conductor, connect these conductors to each other wherever they intersect, and also to each support pedestal they pass or approach under the floor.

These steps minimize the ground planes's impedance at its operating frequencies, and also keep the impedance within the grid both low and equally distributed. At power frequencies any current entering the grid will spread out geometrically at each grid junction, dramatically reducing current density and thus magnetic fields at any single point. At high frequencies, the entire grid will act like one plate of a "spatial capacitor," and a traveling wave injected at one point will thin out, again reducing the effects on equipment in the room.

The grid will act as a broad-band ground plane beneath the electronic equipment and will aid in equalizing potentials between various devices, in reference to "ground," and to the facility itself (i.e., building steel, etc.) The resultant ground plane should be effective at 60 Hz and throughout most of the high-frequency radio spectrum (i.e., to about 30 MHz) without supplemental bonding of the devices being required.

Copper conductor SGR. If the bolted-stringer type of raised floor is not installed, a grid can be created by laying a mesh on the floor slab, under the raised floor. This mesh also can be of copper wire (least desirable), flat copper strap, or braided copper strap (most desirable). The grid must be electrically connected every 2 ft to make a grid of 2-ft squares. Since this grid is about 2 ft farther from the equipment enclosures and requires a longer connecting strap from enclosure to grid, its effectiveness is reduced somewhat.

On new construction, this option in the form of a prefabricated strap grid is possible. It cannot be used for retrofit purposes. The grid could be made of No. 4 AWG copper, also on a 2 × 2-ft pattern, bonded to each column (and support post if applicable) and to each other at each intersection. Note that at the frequencies likely to affect the equipment, most of the current only flows over the outer skin of the conductor. Therefore, the size of the conductor is for mechanical stability, not conductivity.

OTHER ITEMS GROUNDED TO SRG

To be fully effective, all other equipment in the solid-state equipment room should be bonded to the SRG with at least two bonding straps, preferably of rectangular cross section. The SRG acts as a bonding point for all conductors (a wire, conduit, or pipe) that enters or leaves the area. The following items must also be bonded to the grid.

Device boxes enclosing any receptacle connected to the wiring system for the building at large. These receptacles must not be used to supply any of the solid-state devices. Suitable labeling should be provided so everyone will be aware of this restriction.

Electrical, alarm or control panels, whether wall or column mounted. Also, bond all environmental support or control (HVAC) equipment located within the room to the ground plane as well. This includes any metallic floor-level plumbing and any conductive fire-suppression plumbing.

Conductive electrical raceway (or cable assembly at a terminating enclosure) as close as practicable to the point where it enters the area. Vertical penetrations must be bonded to one of the transverse bonding conductors, or to a pedestal of a rigid-grid system. There must be no exceptions to this. Note, however, that loosely jointed jackscrews should not be used as a bonding point.

Cable shields. Digital signals consist of rapid pulses from near zero to 5V (or less on some newer systems) with a specified repetition rate (9600 Hz, 28.8 kHz, etc.) As these signals degrade, the sharp corners of the original square wave become rounded and less distinct. Gradually undetected transitions occur, each one an error. If shielded twisted-pair cables are used, the shield helps control interference.

Note, however, that if these cables run outside the building so as to be subject to lightning exposure or higher voltage crosses, Sec. 725-54(c) applies, which then incorporates numerous rules in Article 800, including Sec. 800-12 on protectors and Sec. 800-33 on cable grounding. Sec. 800-33 is frequently misread because the insulating joint permitted in that section only allows the sheath to be ungrounded inside the building. It does not waive the protector rule. The building must be protected against hazardous voltages that could otherwise enter over a conductive sheath. This is also long-standing Bell System practice.

The protector used is a form of arrester and is not solidly grounded until it closes on a surge, typically at about 60V and then maintaining itself closed down to about 15V.

Since information technology signaling voltages usually don't run much over 5V, single-point shield grounding can be provided with this equipment installed. However, there will be no magnetic field protection unless the cable shield is grounded at both ends, and coupled electric field noise is only attenuated at the grounded end.

So a problem remains: cables have two ends; which end gets the grounding connection and which end floats when they connect two offices? Think about a bidirectional data cable: now which end gets the grounding connection? The answer is that if circulating currents are a problem, consider solidly grounding one end and grounding the other end through a capacitor. This will block the DC and low frequency current while retaining an effective shield at high frequencies.

Surge protection. Lightning and surge protection must be included in an adequate grounding design for these areas. Surge protection must be provided at the service and at all intervening levels of the distribution. Each surge arrester must be bonded to its distribution panelboard or switchboard enclosure, and the grounding return path must be secure, all the way back to the source.

The individual protective devices must be connected to ground through a conductor that is no longer than necessary, and that avoids unnecessary bends, per Sec. 280-12. The perimeter bonding conductor is usually the best reference point for these devices. Never coil the leads, or bend them at sharp right angles. The lightning waveform, although DC, isn't smooth like a battery. Instead it approximates a 100 kHz radio wave, and a conductor with loops and right angles introduces unnecessary impedance. In addition, once it interacts with the impedance of metallic objects in the building, it becomes partially AC in character. Given the levels of current, particularly from a lightning strike, a very small increase in impedance can lead to thousands of volts on the system.

Within the information technology area, the branch circuit supplying each PDU must have surge protection at a junction box located under the floor nearby. The surge protection and the

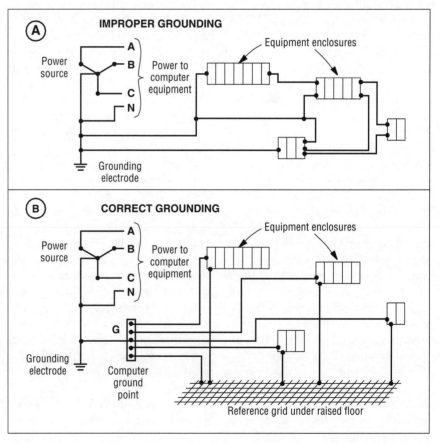

Fig. 6.4. *Computer equipment grounding techniques. (**A**) illustrates grounding that is not appropriate for this type of equipment. (**B**) illustrates proper grounding utilizing both equipment grounding and a reference-grid for high frequencies. Note that all ground leads for grounding of the equipment (not to reference grid) must be run in the same cable or raceway with power conductors. All connections from the equipment to the reference grid must be kept as short as possible.*

junction box must be securely bonded to the grid structure (as shown in **Fig. 6.1**), in at least two places for reliability and low impedance. In addition to the location shown, the protection could be located on the wall at the point of entry, which is often even better. The worst location is right in the PDU.

The typical telecommunications surge protector needs to be supplemented by a high-performance one using solid-state shunt elements that operate just above the permitted signal's maximum voltage. The two types are cascaded. The signal reference grid is a good ground reference for both.

AVOIDING GROUNDING SYSTEM NOISE

Solid-state device grounding must comply with the code and will be safe if it meets the requirements given in Article 250. However, several additional provisions are necessary if the installation is to operate with minimum noise problems from the grounding system. If computer grounding has been installed in the same way as that to most other electrical equipment, the grounding might be like that shown schematically in **Fig. 6.4(A)**.

Here the grounding conductors are run with the power conductors from the source to each piece of equipment, but there are also many pieces of equipment fed from other equipment, and many other metallic interconnections between the grounded metal enclosures of various units. This can be perfectly safe, and in accordance with the code for equipment grounding, but the multiple points of grounding can have slight differences in potential, causing small ground currents to flow—especially under transient conditions. These ground currents appear as noise to the data-processing equipment and can cause errors in computation or component failures. This arrangement produces multiple paths (daisy chains) for the circulation of ground currents (ground loops).

By modifying the grounding as shown schematically in **Fig. 6.4(B)**, these ground loops can be eliminated. Each piece of equipment is fed separately and radially from the source and is grounded (by means of the metallic raceway, green-wire ground, and any other grounding conductors) to a single computer ground point (G) at the source or distribution point. This radial grounding eliminates all ground loops. In addition, each piece of equipment must be connected by the shortest possible connection to the signal reference grid.

While these connections seem to create multiple ground loops, all points on this grounding grid are at the same potential, not only at the power frequency of 60 Hz, but also at much higher frequencies, so no currents will circulate. One point on the SRG must also be tied to the computer ground point to put it at the same ground potential as all other equipment.

A radial installation made as shown to a single ground point can comply in all ways with the NE Code, can be entirely safe, and at the same time can be as free as possible from computer noise problems originating from the grounding system.

GROUNDING ELECTRODES

Grounding electrodes are central to the establishment of a grounding system that meets the safety requirements of the NE Code and operational needs of solid-state electronic equipment. Its primary function is to establish a very low resistance reference to earth potential. This greatly increases the predictability of current flow within the grounding system in a facility. In turn, this permits the development of plans to prevent circulating currents within the grounding path from disturbing sensitive electronic equipment.

As was noted previously, NE Code permits the following to be used as grounding electrodes: a metal underground water pipe; the metal frame of a building; concrete-encased electrodes; a ground ring encircling the building; rod and pipes not less than 8 ft in length; and plates not less than 2 sq ft.

Specific detail on required materials, lengths, sizes of conductors, etc. is given in the NE Code for each of these types of grounding electrodes.

Buried conductors, such as ground ring, are sometimes used as the grounding electrode. It is the preferred grounding electrode for an installation containing extensive solid-state electronic devices.

A No. 2 AWG minimum conductor buried at least 2 ft below grade must encircle the building in order to qualify as a ground ring. The length of the cable will depend on the required soil resistance. Digging the trench to a depth below the frost line of the soil is recommended.

Usually a 100 to 150-ft length of conductor is satisfactory. Sometimes a crowfoot layout utilizing three, four or more runs in various directions from a central point is used. The advantage of crowfooting over a single length is that the crowfoot's shorter lengths will yield a lower total impedance, and this is important.

Made electrodes. If a metal underground water pipe alone is used as the grounding electrode system, the use of an additional electrode to supplement the water piping is required. Typically the additional electrode used is a "made" electrode (a ground rod, plate, or pipe).

A grounding system involving made electrodes consists of the grounding electrode conductor(s), the grounding connections, and the grounding electrodes. Both the grounding electrode conductors and the grounding connections must have an adequate cross section so that they can carry the worst case of ground-fault current without overheating. In addition, the grounding connections must have sufficient mechanical strength to withstand abuse. It is important to note that where a ground clamp will be directly buried, it must be marked for that service, in accordance with applicable testing laboratory standards.

When a driven ground rod or pipe, or a plate is used, either to supplement the metallic water piping system or because no other electrode is available, the NE Code says that the ground resistance cannot be greater than 25 ohms. If greater, then one additional made electrode must be installed at least 6 ft away from the first electrode and connected to it by a suitably sized grounding electrode conductor. This grounding conductor need not be sized greater than a No. 6 AWG copper, or a No. 4 AWG aluminum conductor.

SYSTEM RESISTANCE

Providing the proper ground electrode, however, is only the first step in providing an adequate grounding electrode system. The concept that should be understood is how ground current is transferred from a made grounding electrode to the earth. The resistance of a made grounding electrode is determined by the resistivity of the soil, which is affected by a number of conditions, such as the type of soil, its moisture content, its salt content, and its temperature.

Soil resistivity is defined as the resistance of soil as measured on opposite faces of a soil cube, and is usually given in ohm-inches, ohm-centimeters, or ohm-meters. It is represented by the Greek letter rho (ρ).

Because of its size, the earth is able to absorb an unlimited amount of current and, for that reason, is considered to be an unlimited conductor. But since the earth is composed mostly of silica and alumina (both excellent insulators) along with soluble salts and moisture (excellent conductors), its resistivity varies from 2 ohm-meters to 30,000 ohm-meters or greater.

To determine earth resistance of a rod or cable grounding electrode to earth, the basic equation is:

$$R = (\rho)L/A$$

Where: ρ is the resistivity of the soil; L is the length of the electrode in contact with the earth; and A is the cross-sectional area of current path.

A factor to consider when applying the equation, however, is the variable resistivity of the earth. The equation assumes a constant earth resistivity. It disregards the fact that soil will often gain or lose moisture, gradually change its chemical composition, or change temperature throughout the year.

As the table in **Fig. 6.5** shows, ρ is directly related to moisture content. For example, red clay soil with 10% moisture has a resistivity 30 times greater than the same soil with 20% moisture content.

The chemical composition, or soluble salt content, of soil presents another variable to resistivity, with a reduction in salt content increasing the resistivity.

Finally, as air temperature drops below 32°F, the earth freezes and its resistivity increases rapidly, as shown in **Fig. 6.6**. For example, a drop of 10°F below freezing increases earth resistivity 10 times.

GROUND RODS

A ground rod is the best example that can be used to describe ground resistance and the installation of a made grounding electrode. A ground rod driven into the earth perpendicular to the surface is considered to be surrounded by concentric cylinders or shells of earth with the cross-sectional area of each succeeding cylinder becoming larger as it extends from the electrode (**Fig. 6.7**).

Regardless of soil composition, as the grounded current travels away from the electrode, each successive cylinder of earth encountered by the grounded current becomes larger and larger in cross-sectional area, until a point is reached where, because of the extensive area involved, little resistance is met. Beyond a distance equal to the length of the electrode, the resistance is insignificant. For example, a 10-ft long ground rod makes an electrical connection to an earth cylinder approximately 20 ft in diameter and 10 ft deeper than the electrode. These dimensions constitute the "relative cylinder of earth."

A major commonly committed error is to place the electrodes too close

Moisture content (% by weight)	Resistivity (ohm-cm)	
	Top soil	Sandy loam
0	1000 x 10⁶	1000 x 10⁶
2.5	250,000	150,000
5	165,000	43,000
10	53,000	18,500
15	31,000	10,500
20	12,000	6,300
30	6,400	4,200

Fig. 6.5. Soil resistivity (r) is directly related to the percentage of moisture in the soil.

Temperature		Resistivity (ohm-cm)
°C	°F	
20	68	7,200
10	50	9,900
0	32 (water)	13,800
0	32 (ice)	30,000
-5	23	79,000
-15	14	330,000

Fig. 6.6. Soil resistivity (r) is directly related to the temperature of the soil. The example given is for sandy loam with a 15.2% moisture content.

together, into the same relative cylinder of earth, in effect making several electrical connections to the same conductor without obtaining an effective ground. Thus, two 10-ft rods should be driven a minimum of 20 ft apart to obtain the optimum effect of parallel rods. If two 8-ft rods are used, they should be at least 16 ft apart.

Keep in mind that the diameter of a ground rod has a minimum effect on its resistance. The same relative cylinder of earth is created by both a 1/2-in. and a 1-in. diameter rod. Increasing the diameter of a 10-ft rod from 1/2 to

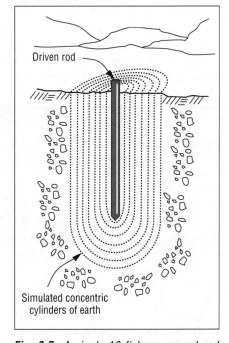

Fig. 6.7. A single 10-ft long ground rod makes an electrical connection to a "relative cylinder of earth," shown by the dashed lines, which has about a 20-ft diameter and is 10-ft deeper than the ground rod or electrode.

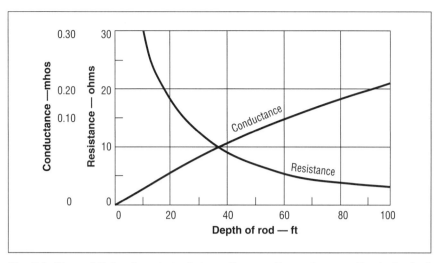

Fig. 6.8. The resistivity of an electrode-to-earth connection varies according to its depth in the ground.

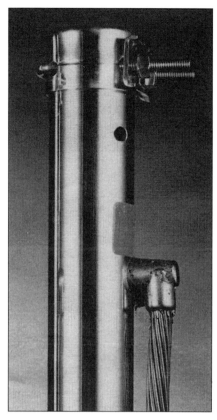

Reduction of soil resistance can be accomplished with specialized electrodes containing metallic salts.

1-in. reduces resistance by only 9½% while increasing the weight by 400%, and increasing the cost considerably. A longer rod should be used to increase the relative cylinder of earth.

Fig. 6.8 shows that, assuming uniform soil resistivity, the resistivity of an electrode-to-earth connection changes with the depth in the ground. This curve ignores seasonal moisture content, changing chemical composition at different soil layers, and frozen soil conditions.

While parallel-driven rods are considered to follow the laws of a parallel circuit (two rods reduce total resistance 50%, three rods reduce it 66%, four rods 75% and so on), this precise relationship is never quite realized.

A guide for calculating the required number of rods is shown in **Fig. 6.9**. As an example, in a substation occupying 1000 sq ft that requires a 5-ohm ground, assume that a single driven 10-ft test ground rod measures 25 ohms. Thus, the ratio of 5 ohms required to a 25-ohm test figure, is a 1 to 5 resistance ratio (a conductance ratio of 5 to 1). Projecting from a conductance ratio of 5 to the curve for 1000 sq ft and then down, the answer is that eight rods are required to yield 5 ohms. This chart also provides the maximum obtainable conductance ratio, which is 6.5 for 1000 sq ft.

REDUCING RESISTANCE

A method called "salting" is sometimes used to reduce the resistance of the soil around a made electrode when deep driving or extensive parallel grounding procedures are not practical because of soil conditions. Salting with metallic salts reduces the resistivity of the soil and thus the total resistance of the relative cylinder of earth around a driven ground rod.

Generally, three different procedures are used to apply the chemicals, as shown in **Fig. 6.10**.

Trench method involves excavating the soil in the shape of a donut about a foot deep around the rod, and then pouring the chemicals into the trench. This procedure eliminates direct contact of the chemicals with the rod.

Basin method involves excavating the earth up to the ground rod and ap-

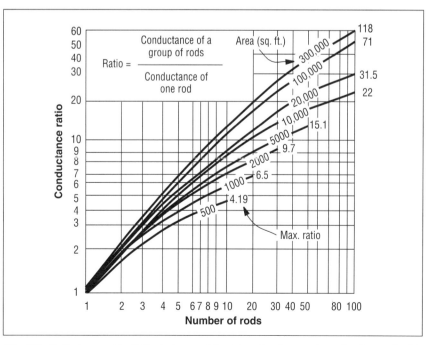

Fig. 6.9. Reference guide that can be used to determine the number of rods required to achieve a given resistance to earth.

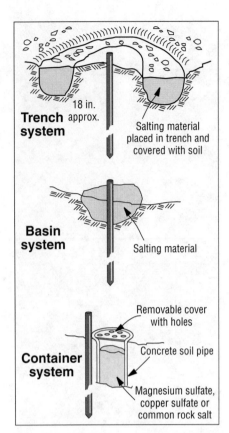

Fig. 6.10. Three dispensing techniques used to reduce earth resistivity by depositing metallic salts in the soil around a driven ground rod.

plying chemicals in contact with the rod. Both of these procedures act rapidly if sufficient rainfall is available to leach or draw the chemicals into the soil.

Container method is a slower acting but longer lasting method. A tile pipe about 16 in. long and 8 in. in diameter is buried 4 to 6 in. away from and parallel to the ground rod. The container is then filled with chemicals and a cover is placed on the tile opening, permitting the chemicals to slowly wash their way into the soil on one side of the electrode.

Specialized version of the container system is a listed grounding system that uses a 2-in. diameter copper tube filled with metallic salts, available in lengths from 8½ ft to 20 ft, or greater. Changes in atmospheric pressure "pump" air through the breather holes at the top of the tube. Moisture in the air condenses on the sides of the tube and trickles slowly down through the bed of metallic salts to the bottom of the tube.

Resistance-reducing powder low in sulfate and chloride is also used for maintaining high conductivity. It is poured around a buried conductor in a trench or an augered hole around a driven ground rod. The powder, which does not depend on the continuous presence of water to maintain its conductivity, absorbs moisture from the surrounding soil and sets to a hardened state.

SURGE AND TRANSIENT PROTECTION

Protecting solid-state electronic equipment from AC powerline disturbances is a growing concern. Computers are subjected to data errors, crashing, and are sometimes damaged or destroyed by voltage transients as a result of an absence or misapplication of protective devices. Other electronic equipment, such as PLCs, solid-state motor controllers, variable-speed drives, communications equipment, and the like, have similar potential for damage from transients, and the protective solutions are almost identical.

Damage from transients can be immediate, where the transient voltage causes a breakdown of solid-state components in the equipment due to the energy of the transient. Damage also can be latent. This is a much more subtle effect. Insulation or components are severely stressed by a single or repeated transients, but not to the point of immediate failure. Each hit weakens their ability to withstand additional stress. At some later time, a transient or other stress that would normally cause no problem occurs at the weakened component, and it fails without any apparent cause.

Fortunately, transient voltage protection is probably less costly than solving any of the many other power problems that affect solid-state electronic loads. When properly applied, transient protection can be extremely effective.

Transients are generally divided into two basic categories: those deriving from natural occurrences (lightning) and those generated through the use of equipment, either on-site or elsewhere. The first line of defense should be at a point that is furthest away from the equipment being protected. This could be at the pole line, if the facility is fed from an overhead line, or at the building service entrance. This line of de-

It does not take a direct lightning strike to cause damage within an electrical distribution system. The electromagnetic field associated with a lightning stroke is responsible for much of the damage that takes place during a storm.

fense, however, must be reinforced by another line of defense; surge arresters located near the items to be protected to intercept transients that penetrated the first line of defense as well as those generated within the facility.

LIGHTNING

Lightning produces extremely powerful, short-duration transients, either by a direct strike or a near "hit." In most instances, damage to sensitive equipment within a building is caused by a lightning strike inducing a surge on local power distribution lines. Research shows that a lightning stroke produces more energy than previously considered possible. A typical lighting stroke carries nearly 3000 million kW at approximately 125 million volts and an average current of over 20,000A!

Lightning strikes also can have secondary effects that cause problems to solid-state equipment in a building. Circuit breakers protecting utility lines can trip and then try to reclose. The resulting voltage sags and outages can cause more problems to computers and other electronic devices than the voltage transients.

The best representation of the wave behavior of lightning-induced energy can be found in a diagram entitled "Lightning Strike", as shown in **Fig. 7.1**,

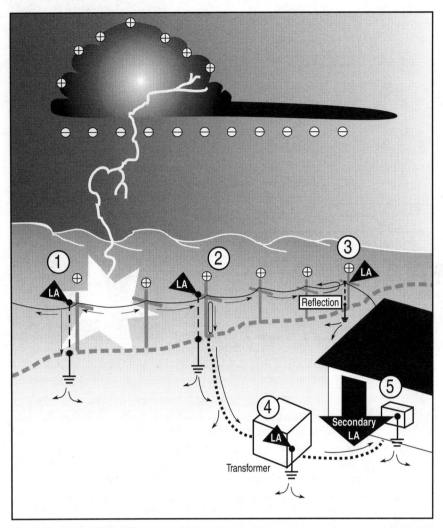

Fig. 7.1. *FIPS PUB 94 "Lightning Strike" diagram shows the wave behavior of lightning-induced energy as it travels along overhead power lines, down pole grounds, and through the earth.*

from FIPS PUB 94. To illustrate what is shown, assume the following:
- The pad-mounted transformer powering the building is rated 13.8kV-480Y/277V;
- it is served by a 13.8kV overhead line, which is being struck by lightning from a thunderstorm cloud; and
- the overhead line is protected with lightning arresters (LAs) rated to spark over at approximately 30kV.

That level of protection provides for the full peak capability of the 13.8kV distribution line while protecting the system from any increased discharge problems. The wiring system and the apparatus connected to it are most likely rated for 95kV BIL. With the LAs operating as required, cross arms, wire and insulation, insulators, and the transformer's primary insulation are safely protected.

While the lightning charge is building up to 30kV on the overhead power line, the resultant increasing energy acts as a traveling wave, traversing the overhead line through Points 1, 2, and 3. At any point where the wave encounters a "discontinuity", it then treats that point as a reflecting point, as indicated by the arrow at Point 3. In reflecting back along the line, the traveling wave follows the laws of physics and doubles its voltage. If it is traveling into the reflecting point at 30kV, it would travel out of and away from the reflecting point at 60kV. This doubling effect takes place at dead-ends, open disconnect switches, or the front end of transformer primaries.

Most power companies protect their dead-ends with LAs. Pairs of LAs, as shown in **Fig. 7.2**, also are placed on either side of in-line disconnect switches, thus providing lightning protection from each direction.

Transformer manufacturers and power companies also make sure that the front end of a transformer primary is protected with surge protection equipment because it also serves as a reflecting point for the traveling energy wave. At Point 4 there may be a doubling of the nominal 30kV spark-over characteristic, even as the LAs on the system operate. When the transformer receives this 60kV surge on the primary side, there is a good possibility that a residual surge will occur on the secondary, in proportion to the turns ratio of the transformer. Thus, the secondary service entrance to a building should also be protected against such a possible residual surge.

Operation of the LAs on the distribution system may require the use of an LA at the service entrance. In other words, a secondary LA at the service entrance should not be thought of as an option that can be included or excluded. It's a protective element that is needed in the system just as much as a station arrester or distribution line arrester on an electric utility system. Only one properly rated and installed unit is required for each building and will protect everything from the basement to the roof. These secondary arresters for applications at 600V and below are, in fact, the third part of the lightning diversion system.

The spark-over energy at an LA located at Point 5 will be close to 1300V, allowing again for the peak voltage of the normal 600V maximum rms wave. This device will normally have a minimum rating of 40kA but is available at higher protective levels, depending upon the application, up to 120kA. With this capability, it's important that such a device be applied at the main service entrance ground, and not somewhere inside the building at a considerably lesser prepared ground application.

On operating voltages of 650V and below, the peak of the normal wave may be near 1000V. Thus, the protection level needed to maintain the shape of

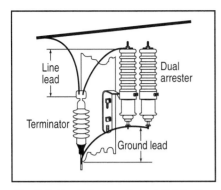

Fig. 7.2. Parallel connection of distribution-class arresters improves protective margins by dividing high surge current discharge through cascade operation.

the sine wave would be slightly higher, say 1200V. Be aware that, with the secondary surge protector doing its job, a transient voltage close to 1200V may travel throughout a building's distribution system. Suppose the voltage transient is only 1150V. Now, the secondary protection will not operate and the 1150V will enter the building unrestricted. In these cases a third level of TVSS protection is needed to lower the level of transient energy. This can take the form of individual devices mounted at subpanels or sensitive loads.

SURGES INDUCED BY EQUIPMENT

Arresters at the service entrance will not necessarily eliminate all powerline surges within a building. Although of greatly reduced magnitude, those that slip by can add to transients that are generated within the facility and cause significant problems to solid-state equipment located there.

Equipment-caused transient voltages result from the basic nature of alternating current. A sudden change in an electrical circuit will cause a transient voltage to be generated due to the stored energy contained in the circuit inductances (L) and capacitances (C). The size and duration of the transient depends on the value of L and C and the waveform applied.

Continuous surges, 250 to 1000V, can come from the operation of electric motors or other inductive loads. Other causes include DC motor drives, the power electronics of variable speed AC drives, DC power supply switching, and even portable power tools.

Momentary surges, 250 to 3000V, can originate from the switching of inductive loads. When an inductor's current is interrupted, a surge voltage is generated. Its magnitude is equal to:

e = L di/dt; the inductance times the rate of current shutdown. These surges can be induced by the opening/closing the contacts of electric motor starters, or the use of arc welders and furnace igniters. When the conductors carrying these surge currents are in proximity to conductors of signaling or data circuits, induced voltages can be generated within these circuits. The result is the introduction of noise and loop currents.

SURGE ARRESTERS

In **Fig. 7.3**, the nonlinear elements (called valve blocks) in most distribution arresters are principally silicon carbide. When a lightning surge reaches the arrester, the gap element arcs over. Sparkover of the gap subjects the valve blocks to a greatly elevated voltage. Because of nonlinear characteristics of the valve blocks (**Fig. 7.4**) their resistance instantly collapses to a low value. The instantaneous surge current, which may be thousands of amperes, flows through the valve blocks to ground. The voltage appearing across the terminals of the protected equipment is directly dependent upon the product of the internal resistance of the arrester and the amount of surge current. The lower this IR discharge voltage, the less the danger to the insulation of the equipment the arrester protects.

Sparkover of the gap elements by a surge discharge ionizes the atmosphere in the arc chambers and a 60-Hz system current will continue to flow through the arrester. This current is known as "power follow current." The high voltage of the lightning surge will have dissipated by this time and the voltage across the valve blocks is approximately the instantaneous line-to-ground voltage of the system. Under such conditions, the valve blocks immediately revert to their original high resistance, with the result that power follow current is reduced to a level that the gap elements in series can interrupt at the next system current zero.

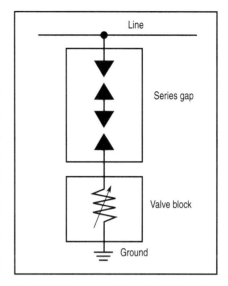

Fig. 7.3. Basic circuit of a distribution-class surge arrester consists of a gap element in series with a nonlinear resistance, silicon-carbide valve block.

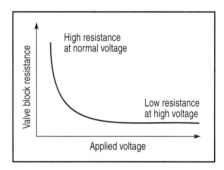

Fig. 7.4. Valve-type arresters depend on the nonlinear-resistance characteristics of their valve blocks for suitable discharge and interrupting capabilities.

There are three valve arrester classes designated by ANSI Standard C62.1:
- distribution class;
- intermediate class; and
- station class.

Many important variations exist in each class that are related specifically to application requirements and voltage ratings.

LOCATING SURGE ARRESTERS

Some users of equipment containing solid-state devices believe in using only lightning protection devices applied at the service entrance of a building. Others believe that individual plug-in transient voltage surge suppressors (TVSSs) at individual loads is the best

Surge arresters are often located at the incoming line compartment of MV transformers. Here they are mounted on the back wall behind the cable terminators.

way to go. Both of these approaches, when taken separately, are open to possible failure. As was discussed previously, proper surge protection is a combination of both approaches. A complete protection system removes transient energy at several points of its travel path, before the energy can exceed the damage threshold of each type of equipment.

Overhead lines. Arresters of proper voltage rating installed along the length of a poleline discharge the lightning surge at some level safely below the BIL rating of the cable it protects. In the performance of their duty, they introduce a traveling wave voltage in the cable equal to either the sparkover voltage of the arrester or the IR discharge voltage of the arrester plus the inductive drop of its line and ground leads, whichever is greater.

Pole risers. The function of a riser-pole arrester installed before the service conductors enter the building is to significantly reduce this surge voltage. A distribution-class arrester is frequently used for this purpose. They must have the shortest possible line and ground leads to be effective. Improved protective margins can be achieved by using intermediate-class arresters.

Underground lines. Of even greater concern as far as lightning affecting solid-state electronic devices within a building is the practice of running power and communications/control circuits underground. It would seem that this would reduce the lightning-induced surge problem. In practice, this is not necessarily the case for two reasons:
• the earth is a conductor; and
• use of nonmetallic conduit for protection of circuits run underground.

When rigid metallic conduit in a concrete ductbank is used for underground circuits, its effect is to shield the sensitive equipment circuits from electromagnetic fields. Nonmetallic conduit does not provide this protection, thus, these circuits are more susceptible.

For underground lines, it is hard to visualize a direct lightning strike. But a phenomenon similar to the one affecting overhead lines takes place. As a consequence of a lightning discharge, electromagnetic fields are induced not only on the surface of the earth but also in the earth. These fields, currents, and voltages in the earth diminish roughly as the square of the distance from the strike, and the resistance of the earth. This rapidly fluctuating field in the earth induces voltages and currents in nearby objects such as underground cables. In turn, these induced voltages and currents create another electromagnetic field that opposes the field created by the lightning. A side flash of very high voltage then occurs to try to equalize the potentials of the fields, thereby introducing noise into the underground circuit, and possible ground loops that can corrupt communications/data circuit signals.

This noise induced into the electronic circuits requires that a further level of surge protection be provided.

SIZING LIGHTNING SURGE ARRESTERS

Modern distribution systems feeding a facility have significant amounts of solid-state electronic loads require effective protection against lightning surges. Where the service is from an overhead line, surge protection is provided on the riser pole carrying the conductors from the overhead distribution system into the service-entrance equipment. This is the first line of defense against surges that could affect computers and other similar devices.

Ratings. As cable ages, its basic insulation level (BIL) can decrease, making it more susceptible to overvoltage. It is important to optimize the cable protection to avoid insulation breakdown and arcing, which can induce noise into the internal distribution system.

The rating of the required protection is not determined simply by matching the BIL rating with the arrester discharge rating. The phenomenon of "voltage-doubling" must also be taken into account. As described before, when a traveling wave reaches a point of high resistance (for instance, an open switch in a circuit), it reflects back just like an ocean wave hitting a seawall. The wave amplitude will approximately double at this point of reflection. Therefore, the value used in calculations must be twice the arrester discharge voltage.

Protective margin. PM is a measure of the ability of a surge arrester to protect a piece of equipment or a system. The calculation involves the BIL withstand rating of the item to be protected (in this case the cable) and the arrester discharge level (often listed in surge-arrester data as its "10kA IR discharge voltage"). The PM can be calculated as illustrated in the following example.

EXAMPLE: Assume a cable with a 125kV BIL rating is employed on a 24.9kV system. A typical silicon-carbide (SiC) arrester for this voltage level would have a listed discharge level of 67kV. Is the rating adequate?

ANSWER: At first glance, the arrester seems adequate. However, if voltage-doubling occurs within the cable, the surge voltage could rise to twice 67kV, or 134kV, which is 9kV higher than the cable BIL. The protective margin (PM) is obtained by dividing this 9kV by the item to be protected's BIL rating.

$$PM = 9kv \div 125kV = 0.072 \approx 7\%$$

Since this is 7% higher than the cable BIL, it is expressed as -7%. This is the protective margin of the installation. Because of the negative value, the arrester in this example is not satisfactory.

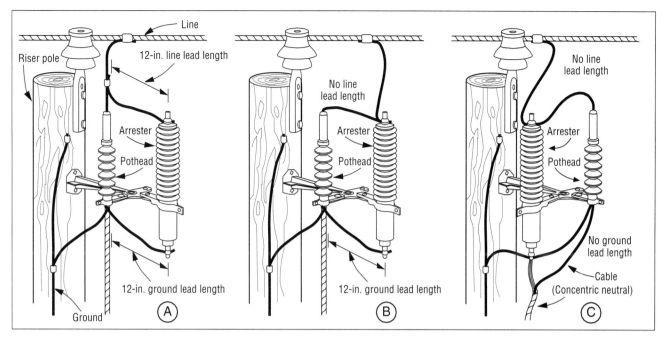

Fig. 7.5. Mounting arrangement shown in (**A**) of surge arrester and cable terminator on riser pole results in a 24-in. arrester lead length. Connection (**B**) eliminates the 12-in. line lead length, leaving only a 12-in. ground lead length. An effectively zero lead length installation is created by connecting both line and ground directly to the arrester (**C**).

A modern metal-oxide (MOV) arrester has lower and more consistent discharge protective levels than old-style SiC arresters. A typical MOV distribution arrester for this application could have a 10kA IR discharge voltage of 59kV.

A typical MOV pole-riser arrester, which is a higher-quality distribution arrester, could have a 10kA IR discharge voltage of 52kV. These two types would provide protective margins of 6% and 20% respectively when voltage-doubling occurs. Both would afford adequate protection.

BIL margin percentages are calculated assuming the industry standard 8 × 20-microsecond wave shape (reaches crest in 8 microseconds and declines to one-half of crest in 20 microseconds). The actual protective margins offered by an arrester will thus vary from the calculated values, depending upon the actual wave shape of the lightning surge (lightning surges much faster than 8 × 20 are thought to be common).

LIGHTNING SURGE ARRESTER LEAD LENGTH

An important consideration in designing the connections between the surge arresters and the cable terminators is the length of conductors used to make the connections, or "lead length." Here is a guideline that can be applied when designing such an installation.

Protective margins calculated from arrester catalog data do not include allowances for lead length. Arrester lead is the combined line and ground lead length in series with the arrester and in parallel with the device being protected.

For example, in **Fig. 7.5(A)**, the arrester is mounted near the cable terminator with approximately 12 inches of line lead and 12 inches of ground lead, for a total arrester lead length of 24 inches.

The commonly accepted method of determining the effect of this lead length on arrester operation has been to use an inductance of 0.4 microhenries per ft and a lightning current rate-of-rise of 4000A per microsecond to determine the L di/dt voltage drop of the installation. Using this method, calculations show 1.6kV per ft of arrester lead length will be added to the discharge voltage of the arrester. This increase in discharge voltage decreases the amount of protective margin on the system.

For example, the effective 10kV IR discharge voltage of a riser-pole arrester would be increased from 52kV to 55.2kV with the addition of 24 inches of lead length. Although this increase may appear insignificant, it is critical when the voltage-doubling effect is considered. Voltage doubling will reduce the BIL protective margin from 20% to less than 14%. Therefore, the arrester lead length must be kept as short as physically possible in all arrester installations, particularly in critical situations such as cable protection.

The installation shown in **Fig. 7.5(B)** is similar except that the line lead is taken directly to the arrester before going to the cable terminator. This installation virtually eliminates all arrester lead on the line side since, by definition, no lead is in series with the arrester while being in parallel with the terminator. Total lead length is reduced to 12 inches (the length of the ground lead).

Fig. 7.5(C) carries the length reduction one step further. The arrester is mounted between the cable terminator and the pole ground. This arrester installation has zero line and zero ground lead length. Therefore, lead length has no detrimental effect on the protective characteristics of this installation.

PROTECTING SOLID-STATE DEVICES

Now consider the needs of sensitive equipment inside a building. There is a difference between a product that diverts as much as 40kA or more into a ground location and one that merely "snips and clips" the high transient voltage appearing on the system. The former is a large "ball of fire" protector while the latter is a high spike penetration protector. Note that the operation of the service entrance protector itself may create a problem inside the building, thus requiring "snip and clip" transient protectors downstream of the main service entrance protector.

For example, consider a wave that exceeds the 1300V level of the secondary protector at the service entrance ground. This protector will then spark over and conduct the energy in excess of the 1300V into the earth. Meanwhile, there is still 1300V left on the system and traveling into the building. Now there is a need to remove the damaging effects of this residual transient energy from the rest of the building's internal system. While the 1300V spike is not a large "ball of fire", it is, nevertheless, a very large penetrating transient for sensitive electronic devices. Therefore, this residual must be taken off the line as quickly as possible so that it will not penetrate sensitive chips or disturb other parts of the system.

In large buildings, this equipment protection may be done more economically with a panel-style protector for all solid-state circuits fed from the panelboard. In smaller buildings where no subpanels exist, it may be required to go directly to discreet equipment protectors, which couple together the protection of both power and communication/data circuits.

TRANSIENT SUPPRESSORS

Stopping transients from reaching sensitive electronic equipment is accomplished by the installation of a transient voltage surge suppressor (TVSS). Secondary surge protectors are installed near the sensitive equipment to protect it from residual voltage surges that are permitted to enter the system when a lightning arrester operates, and against transients generated within the facility. There are several types of TVSS devices.

Crowbar devices include air gaps, gas-discharge tubes, lightning arresters, and switching devices (**Fig. 7.6**). They

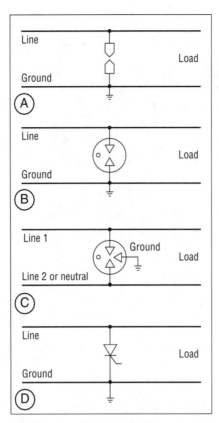

Fig. 7.6. Crowbar devices: air-gap carbon block or metal electrodes (A); two-electrode gas discharge tube (B); three-electrode gas discharge tube (C); and thyristor switching circuit (D).

are inactive until the voltage applied exceeds the rated or set value. At that point, they become conducting, creating essentially a short-circuit path to ground and diverting the high transient voltage away from the load. Most lightning arresters, at all voltages, are crowbar type. When these devices conduct, the normal voltage on the system drops as a result of the virtual short-circuit to ground. When the overvoltage is cleared and the protective device ceases to conduct, normal voltage is restored to the system.

Gap-type devices require between 300 and 700V to initiate an arc across the gap. Switching devices, such as thyristors, are gated on by a control circuit. Both types, during the time they are conducting, must carry not only the energy of the transient but also the short-circuit current (follow current) of the system that is shorted to ground for durations up to one-half cycle (8.33 ms) or longer. The arc in the gap ceases to carry current at current zero. At the first or a later current zero, the energy from the short-duration transient has been dissipated and the system voltage is not high enough to cause a restrike; the nonconductive gap is restored.

Switching devices cease to conduct at current zero, and there is no overvoltage to cause it to be gated on again. If the follow current exceeds the rating of the device, interruption may not take place at the first current zero, and some type of backup protection may be necessary.

Gas-discharge tubes (GDTs) are the most commonly applied gap devices for computer systems and other sensitive equipment. They consist of two or three electrodes sealed in a glass tube filled with a mixture of gases at low pressure. The electrodes form one or two metallic gaps. GDTs have a relatively long life and high current-carrying ability.

Carbon-block spark gaps are common in low-voltage telephone and data systems. They consist of two carbon-block electrodes separated by an air gap. This type of device is low in cost, but has a comparatively small current-handling ability and short life. Spark-over voltage for a given unit is not constant and can range from 300 to 1000V, unpredictably.

Crowbar devices are relatively slow, taking several microseconds to operate, subjecting the protected system to the very fast initial voltage rise, which can be several thousand volts per microsecond. Additional surge protection, therefore, must be used downstream to protect equipment against this surge.

Voltage clamping devices such as varistors (nonlinear resistors), metal-oxide varistors (MOVs), zener (avalanche) diodes, and selenium rectifiers, are unidirectional conductors until a breakdown voltage is reached, at which time they conduct in the reverse direction. They are normally connected in the circuit in the nonconducting direction, presenting a very high impedance. At the clamping (maximum permitted) voltage, somewhat above normal line voltage, the impedance drops rapidly. The higher the current, the lower the impedance becomes, shunting the transient away from the load.

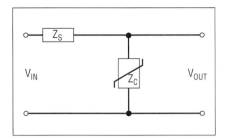

Fig. 7.7. Voltage divider action of a clamping device.

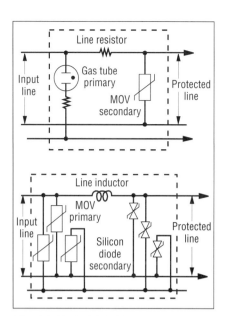

Fig. 7.8. Typical hybrid surge arrestors.

Clamping-device impedance, in series with the source impedance, acts as a voltage divider and maintains the system at or near the clamping voltage (**Fig. 7.7**). The relationship is:

$$V_{out} = V_{in}[Z_C \div (Z_C + Z_S)].$$

When V_{in} is normal, Z_C is very high compared to Z_S and:
$[Z_C \div (Z_C + Z_S)] \approx 1$; thus, $V_{out} \approx V_{in}$

When V_{in} is high (transient above clamping voltage), Z_C is also low and of the same magnitude as Z_S. As V_{in} increases, Z_C decreases and the current through Z_C increases. As a result, the output voltage (V_{out}) remains relatively constant at the clamping voltage.

At the clamping voltage, the clamping-device impedance goes low, but it does not fall to nearly zero, as occurs with crowbar devices. There is, therefore, essentially no short-circuit magnitude follow current. Clamping voltage is maintained on the system load. The lower the source impedance, however, the more difficult it is for the clamping device to maintain system voltage without drawing excessive current.

Clamping devices operate in the nanosecond range, thousands of times faster than crowbar devices, but are generally capable of dissipating considerably less transient energy. They effectively clamp voltage spikes to a maximum level but have no effect on notches or brief dropouts.

MOVs are available for a wide range of voltages and currents, from as low as 4V for data lines to several thousand volts for power systems; for peak pulse currents from only a few amperes up to tens of thousands of amperes; and for energy dissipation from less than 1 to over 10,000 Joules. They are low in cost, compact in size, easy to apply, and are among the most frequently used devices for transient protection. One disadvantage of MOVs is that their characteristics deteriorate with repeated transients. Although this problem is being reduced with improved technology, MOV life under frequent transient conditions must be considered.

Zener diodes also are used to provide voltage clamping. These are avalanche devices that maintain a high impedance until the breakdown voltage, and then rapidly drop in impedance to provide the clamping action. Zener diodes used for clamping have different characteristics from those used to provide voltage regulation. Under avalanche conditions, when the diode is clearing an overvoltage, substantial current must flow across a thin semiconductor junction. The ability to dissipate heat produced at the junction limits the maximum energy that zener diodes can handle. Zener diodes are faster acting than MOVs and provide very effective clamping, to a nearly constant voltage, but with limited energy-handling ability.

Silicon-carbide varistors have high power-handling capability and are used in high-voltage surge (lightning) arresters. They tend to draw considerable current in the normal state, so they are commonly used in series with a gap that provides an open circuit until a surge occurs. This property makes them unsatisfactory for low-voltage clamping operation. Some newer zinc-oxide lightning arresters have better nonlinear characteristics and can be used without a gap. They are essentially crowbar devices, but perform almost like clamping devices.

Selenium-cell transient suppressors, of specially designed selenium rectifier elements, are occasionally used. They can handle high energy transients and repeated hits very well, but have poor voltage-clamping ability. Combination units are available, however, that uses selenium devices, custom-made to minimize the undesirable characteristics and take advantage of the desirable ones.

Hybrid devices (**Fig. 7.8**) are transient suppressors that combine two or more technologies to provide transient suppression over a wider range of voltages, rates of rise, and energy content, than any one device could do successfully. The most common hybrid combines a three-electrode gas-tube crowbar, for high-energy capability, with MOVs or zener diodes for nanosecond response. The clamping devices take care of the initial voltage rise passed through by the slower gas tube and transients too small to activate the gas tube. Sometimes, a filter is added to provide further attenuation, especially for high-frequency oscillations.

Surge reference equalizer. It has been found that with multiport loads (such as TVs with AC power and CATV ports, or fax machines with AC power and telephone ports), a transient voltage surge on one port, even if protected by a TVSS, causes a surge to be impressed across the other ports, often causing damage to the load equipment. One potential solution is the use of a surge reference equalizer to prevent differences in the ports' ground rererences under transient voltage surge conditions. It is a device used for connecting multiport equipment to external systems. All conductors connected to the load to be protected are routed, physically and electrically through a single enclosure with a shared reference point between the input and output ports

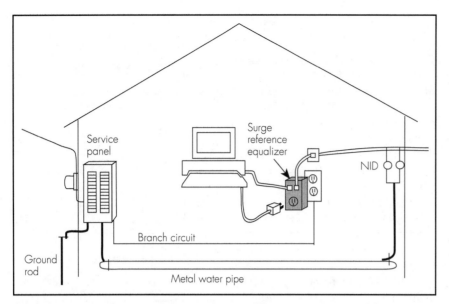

Fig. 7.9. A surge reference equalizer can prevent differences in computer AC power input and data communication ports' ground references under transient voltage surge conditions.

of each system, as shown in **Fig. 7.9**.

ATTENUATION DEVICES

Attenuation devices are inserted in a circuit to permit power at line frequency to pass, while attenuating transients that are in the kHz to MHz range, several orders of magnitude higher than the 60-Hz power.

Low-pass filters are made up of combinations of shunt capacitors and series inductances. Its components must be able to withstand the high transient voltages, and the series inductances must carry the sometimes large continuous-load currents. The filters must be matched to the type of transient to be attenuated. Also, their impedances must be matched to the input and output impedances; since these vary, the filter must make compromises that may do more harm than good. Since they are LC circuits, a transient can send them into oscillation, and the output of badly matched filters can contain more spurious voltages and frequencies than the input.

Active tracking filters are special devices that sense the instantaneous sine-wave voltage at any point in the cycle and, as clamping devices, limit the maximum voltage. This limits the deviation from the true instantaneous sine-wave voltage at any point on the sine wave to ±2V. When a voltage deviation is sensed, the unit switches on to provide full filtering in less than 5 nanoseconds. Not only are spikes clipped, but notches also are "filled in" by energy stored in the filter's capacitance, which cannot be done by clamping devices.

APPLYING TVSS PROTECTION

The first step in protecting computers and sensitive electronic equipment against transients is good installation practice. Careful grounding is necessary, with equipotential grounding for a computer room. With poor grounding, the action of transient suppressors, which divert the surge energy to ground, could cause differences in ground potentials that might create as many problems as the original surge.

Separate protection for individual units usually must be provided (unless already built into the equipment) for distributed control in an industrial plant, PLCs, personal computers, minicomputers, terminals, and the like. Data and communication lines entering from outside the building, and those entirely within the building that may be subjected to transients, must also have individual transient protection.

Incoming power lines should be protected at or near the service entrance by a crowbar device, such as a gas tube or one of the newer fast-acting, low-voltage zinc-oxide lightning arresters. They will protect against the high-energy lightning and switching transients that may come into the building from the utility or as a result of lightning. This device should be rated for the incoming voltage. **Fig. 7.10** is an oscilloscope image of the clamping action of an MOV transient suppressor.

Incoming data lines also should be protected by a crowbar device where they enter the building, and it should be rated for the low voltage of the data line. Telephone lines usually will be protected by the telephone company, probably with a carbon-block gap arrester.

ANSI/IEEE Standard C26.41 and IEC Standard 664 have attempted to define the severity of exposure to lightning transients by categories relating to the location in the building. While the standards differ somewhat, the general consensus, not surprisingly, is that the most severe transients occur outside the building on the service conductors and inside at the service-entrance conductors. Next in severity are those transients at distribution equipment, large

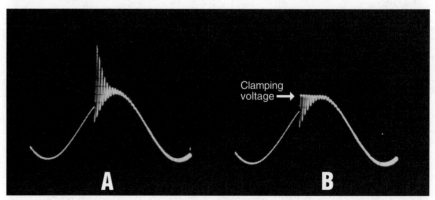

Fig. 7.10. Clamping action of an MOV. An oscillatory transient is shown in (**A**); and the same transient after being clamped is seen in (**B**). Note that the MOV has no effect on that portion of the oscillations falling below the clamping voltage. The MOV limits the maximum voltage but does not filter out the high frequencies.

Location category ANSI/IEEE 62.41	Comparable to IEC 664 category	Impulse		Type of specimen or load circuit	Energy (joules) deposited in a suppressor with clamping voltage of	
		Waveform	Medium exposure amplitude		500V	1000V
A. Long branch circuits and outlets	II	0.5µs — 100kHz	6kV 200A	High impedance Low impedance	(120V system) — 0.8	(240V system) — 1.6
B. Major feeders short branch circuits, and load center	III	1.2/50µs 8/20µs	6kV 3kA	High impedance Low impedance	— 40	— 80
		0.5µs — 100kHz	6kV 500A	High impedance Low impedance	— 2	— 4

Fig. 7.11. Representative surge voltages and currents. In this table, the voltage shown for high-impedance specimens represents the surge voltage; for low-impedance specimens, the current shown represents the discharge current of the surge (not the short-circuit current of the power system.)

buses and feeders, and short branch circuits (near the service). Lowest in severity are those on long branch circuits (**Fig. 7.11**).

Information-technology-room power feeders supplying computer equipment and other solid-state electronics (not HACR or other nonsensitive peripheral equipment) should be protected by a voltage-clamping device where they enter the computer room. This unit should have a fairly large energy-dissipation rating.

ITE-room distribution equipment, whether just a panelboard, or preferably the panelboard on the secondary of a shielded isolation transformer, separate or as part of a power distribution unit (PDU), should be protected by a voltage-clamping device on the mains. It is not necessary to protect each individual branch circuit. This device requires only a moderate energy- dissipation rating.

ITE-room data lines entering require a low-voltage clamping device. Each data line should be protected separately at the point where it enters the room. The protection should be at the low-to-moderate energy level.

Individual solid-state electronic equipment (PCs, terminals, mainframe computers, and similar equipment) should be protected by at least a clamping device, with energy dissipation appropriate to its location. A hybrid device, perhaps with attenuating filters, might be desirable.

SELECTING TVSS EQUIPMENT

Selecting the clamping device for each application is partly a science and partly an art. The device should have a minimum continuous voltage rating that is higher than the system or data line rms voltage, or DC phase-to-phase or phase-to-ground voltage. The rating should be not less than 110% of the connected voltage; and 125 to 150% would be better to allow for system voltage deviations. Selecting the voltage rating is uncomplicated.

Selecting the energy dissipation rating required is less precise and more difficult. It is rarely possible to predict the energy content of a transient that might occur. As the energy rating of a clamping device goes up, so does its cost; therefore, picking an unnecessarily high rating will result in needlessly high cost. The device selected must survive the worst probable transient; and at the same time, the clamping voltage must not exceed the withstand abilities of the equipment being protected.

All gap-type devices have the characteristic that the voltage at which the gap breaks down and arcs is not consistent; although gas-discharge tubes are considerably less erratic than air gaps. When protecting both conductors of a two-wire circuit or data line, it is best to use a three-electrode gas tube that will shunt both lines to ground simultaneously. Two 2-electrode gap devices, one on each conductor, could have different flashover voltages so that one would conduct first. This unbalanced breakdown could cause transient current to flow through the load, causing damage.

Transient suppressors must be installed using the shortest possible leads or conductors, so that the lead inductive reactance is small at high transient frequencies.

PLC controls, *PCs, instrumentation, communications, and similar devices keep production lines humming. Loss of power to these items or contamination of signals will have a significant economic impact, justifying the installation of sophisticated power conditioning equipment.*

POWER CONDITIONING AND RELIABILITY

Selecting equipment to condition power fed to computers and other solid-state electronic loads requires first determining if the system to be protected is critical so that no downtime can be tolerated. Evaluating the site history of power outages and the cost per hour of downtime, against the cost of protective equipment is also necessary. The subject of monitoring and analysis is covered in Chapter 9 of this book.

Power conditioning devices and equipment to improve power quality, in approximate order of increasing cost and complexity, include: noise filters, voltage regulators, shielded isolation transformers, motor-generator (MG) sets, magnetic synthesizers, and uninterruptible power supplies (UPS).

Power conditioners can include a combination of individual types of conditioning equipment. They can contain the voltage regulation, transient protection, noise filters, and shielded isolation transformers necessary to provide output power fully protected against all input power disturbances except sustained outages.

Transient suppressers should also be included in this list, being the primary line of defense against noise that can affect solid-state devices. This subject, however, is so important that it was given its own chapter in this book (Chapter 7).

Noise filters also qualify as power conditioning tools because they remove unwanted signals from the line. Low-pass filters are made up of combinations of shunt capacitors and series inductances. Their use to trap harmonics was covered in depth in Chapter 3 of this book.

VOLTAGE REGULATORS AND LINE CONDITIONER

An examination of power conditioning begins by considering voltage regulating equipment and equipment called "line conditioners." This latter term is generally accepted to mean the combination of a voltage regulator with a noise rejection system, such as a shielded isolation transformer, for pro-

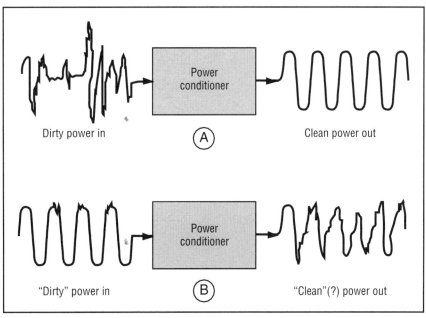

8.1. General understanding of the topic of power conditioning assumes that "dirty" incoming power is transformed by the power conditioning equipment into "clean" power (**A**). In reality, the situation is often that shown in (**B**), where a moderate input waveshape becomes a distorted waveshape at the output of the power conditioner.

tection against common-mode noise and the enhancement of voltage stabilization.

Products in this category (voltage regulators and line conditioners) were among the first recommended by manufacturers of equipment containing solid-state devices to assist in protecting data-processing systems from outside power influences. In the early days of the applications, little consideration was given to load/source interactions. As shown in **Fig. 8.1**, in some cases the resulting conditioned power waveshape was worse at the output than the power source waveshape at the conditioner's input. In time, it became apparent to manufacturers and end users alike that just another box in the circuit might not be necessary, even though it appeared to give additional protection.

Consider that almost all sensitive power supplies used in digital electronic systems are designed with the finest voltage regulator available, one having a range of operation of almost 20% from the top of the voltage limit to the bottom. Also, it responds extremely fast. The ITIC (CBEMA) curve indicates a +6% to -13% range for many of these devices. If these units have such a fine regulator, then why the need for an additional voltage regulation product in line with these units? The answer is that the voltage protection concept has been oversold. Granted, these voltage protection products are the least expensive way of providing some form of buffering ahead of the digital system.

CONSTANT-VOLTAGE TRANSFORMERS

The oldest form of regulation is a type of transformer that supplies a stable output voltage through all current loading within the rating of the product. This is ferroresonant technology known as a constant-voltage transformer (CVT), sometimes referred to as a "saturable magnetic device." The general outline of this concept is shown in **Fig. 8.2**.

This transformer has a core intended to saturate; in other words, it accepts energy into the core, flooding the iron core in such a way that the result is a flat voltage stable across the range of load currents. Within the ratings of the device, the voltage will remain level from light load to heavy load. This is a simple and very good older technology that's still in use today. And it's very economical.

A note of caution: This product is based on sine wave load characteristics dating to a time when harmonic interaction was not yet upon us. As such, it should not be applied in the face of a high harmonic load spectrum without considerable oversizing or design modifications by the manufacturer to assure harmonic content compensation.

Also remember there is an upper limit to the load range of this device; if that limit is exceeded, this product may actually reduce its output slowly to zero in order to self-protect the circuit. This performance trait, known as *current limiting effect*, can be avoided by sizing the maximum load approximately 20% below the full-load rating of the regulator. This type of oversizing also will help in handling harmonics if they're present and haven't been compensated for by the manufacturer.

Finally, remember that because a CVT uses energy to saturate its core, its efficiency will be low if loading is very light. Normal efficiencies for fully loaded units will be in the high 80% area up to 92%, while efficiency under light loads might drop to 45% to 55%. By sizing the load somewhere in the middle, it will be able to operate under most conditions, giving up some efficiency points to an average of 75% to 80%.

TAP-SWITCHING REGULATORS

A more modern version of a transformer for voltage stabilizing is a tap-switching regulator shown in **Fig. 8.3**.

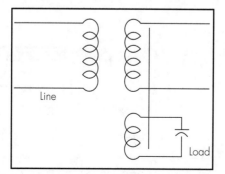

Fig. 8.2. A constant voltage transformer is designed to supply a stable output voltage through all current loadings.

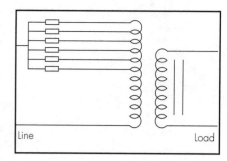

Fig. 8.3. A more modern voltage stabilizing method is a tap-switching regulator, which uses diodes and transistors to perform automatic "stepping" between the taps by sensing the output of the transformer.

This device uses the same concept as a constant-voltage transformer, but in a fully adjustable configuration with diodes and transistors, to perform automatic "stepping" between the taps by sensing the output of the transformer.

Initially, tap switching was done on the secondary; however, this method produced electrical noise. As a result, the design was changed so that the switching is done on the primary; this design remains today and uses a nominal six tap settings.

The unit is based on a regular magnetic core rather than a saturable one, and thus has all the properties of a good isolation transformer. It has excellent overload capability, like a regular dry-type transformer, and has the same low internal impedance. This makes it suitable for interfacing between a high impedance source and a high impedance load; and it does not introduce harmonic distortion to the system.

A tap-switching regulator uses sensing of the output voltage to signal the primary tap change to adjust for the variances in input supply. It can make those adjustments at the current zero crossing in order to avoid electrical noise, and it has a response time of three to five cycles. While this speed of response seems slow by comparison with the speeds of computers, it's just this difference in speed that commends this technology. If this product is used in a circuit where there is a "fast" regulator inside the CPU, then the slower regulator will not interact with the higher speed product; the two will "handshake" well together.

As discussed previously, the effi-

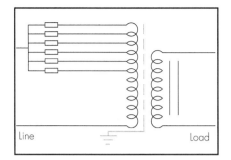

Fig. 8.4. A tap-switching regulator is also available with a shield.

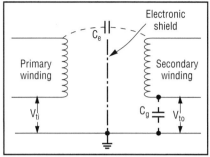

Fig. 8.5. A basic shielded isolation transformer has a grounded metal foil inserted between the primary and secondary windings.

V_{ti}	V_{to}	Ratio	Common mode noise attenuation (db)
5	1	5:1	14
20	1	20:1	26
50	1	50:1	34
100	1	100:1	40
150	1	150:1	44
200	1	200:1	46
250	1	250:1	48
300	1	300:1	50

Fig. 8.6. Common-mode noise attenuation can be described either as a ratio or in db.

ciency of the saturable product falls off rapidly at partial load ratings and requires an assessment of loading for proper sizing. Efficiency of the tap-switching regulator is normally higher and remain high from zero to full load, as one would expect of a standard dry-type transformer.

LINE CONDITIONERS

In both cases outlined above, an electrostatic shield can be added to make the regulator into a combination product. This is one of the best application moves that can be made when a voltage regulator is required in the circuit. With the shield, as seen in **Fig. 8.4**, the product becomes a common-mode noise rejecter and a voltage stabilizer.

A form of voltage regulator, combining stabilization with the supply of harmonic current, is known as an *active power line conditioner*. This technology, which was developed through the Electric Power Research Institute (EPRI), has the promise of a further "combination" solution for the harmonic-requiring load.

SHIELDED ISOLATION TRANSFORMERS

Shielded isolation transformers are not true power conditioning devices but rather "noise rejecters" that provide good common-mode noise isolation. They can be 1:1 ratio, for isolation only, or can also provide voltage transformation. The minimum shielding (**Fig. 8.5**) is usually a grounded single Faraday shield between windings, but additional attenuation can be provided by also shielding either or both individual windings. Common mode noise attenuation (Vti/Vto) achieved is expressed either as a ratio or in decibels (db). The table of **Fig. 8.6** shows the relationship of ratio to db.

In a shielded isolation transformer, capacitance coupling between the primary and secondary windings of an unshielded transformer is replaced by the primary-to-shield capacitance, and that of the shield-to-secondary that results from the small electrostatic field around the shield. The primary-to-shield capacitance conducts most of the incoming transient energy (Vti) to ground. Thus, the effective capacitance (Ce) between primary and secondary windings of a shielded isolation transformer is much less than that of an unshielded unit.

Attenuation of 120 dB or more is said to be required for the most sensitive circuits. For most systems, however, isolation of 40 to 60 dB is usually adequate and realistically attainable.

Characteristics. Shielded isolation transformers cannot adjust their voltage output (Vto) for sags or surges on the input line. Nor will they protect against voltage spikes of subcycle duration and, therefore, surge suppressors should be added to the system. A suppressor will clamp the voltage at preset values (slightly above nominal) and conducts the excess voltage to ground. These devices typically act within 25 milliseconds.

Application. When applying a shielded isolation transformer to effectively protect critical loads such as computers, electronic instrumentation, or data processing peripherals, be sure to locate the unit as close as possible to the load panelboard. Make all ground connections radially, and bond the neutral and ground together at the "newly derived" neutral and attach it to ground according to the requirements of the NE Code.

A common use for an isolation transformer is in a power distribution unit (PDU). Also called a "computer power center," these units combine power isolation with power distribution and permits installation directly in computer rooms.

MG SETS

Don't discount old power conditioning concepts such as the use of motor generators. The MG set (**Fig. 8.7**) with a few modifications and improvements, has developed into a significant and

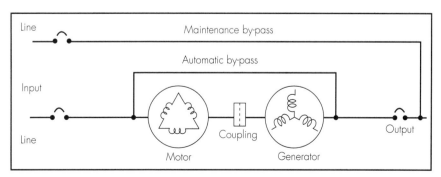

Fig. 8.7. An MG set provides a true mechanical separation between input and output, via the shaft or belt coupling. As a result, both normal-mode and common-mode noise are eliminated.

This coupled in-line MG set functions as a line isolator/stabilizer.

reliable power conditioning device.

The most cited reason for frustration with the "old-style" MG set is bearing failure. The reason is that the majority of units used a two-bearing construction on a common shaft; one bearing on the motor side and one on the generator side. This "balanced" design performed well in applications such as changing frequency, changing from AC to DC or viceversa, or from single phase to 3-phase.

With the coming of age of the computer, however, various changes had to be made to the old-style MG set to enhance its performance and reliability. For one thing, the bearings have been beefed up. In some cases, provisions have been made for multiple sets of bearings: two on either side of the motor and another two on either side of the generator. Even with a bearing failure, the machine can continue to operate on the three remaining bearings.

In addition to improved bearings, more choices of the driving motor are now available.

Induction motor. One chronic problem occurs when an induction motor is used as the prime mover. A characteristic known as slip occurs because the motor rotates at a speed somewhat slower than the synchronous rotation of the generator; for example, 1740 rpm versus 1800 rpm. Thus, the generator has a difficult time keeping 60 Hz frequency under varying input conditions. Also, this slippage increases if the input power is reduced due to voltage decay or temporary loss of power. The solution to holding frequency steady under these conditions is the addition of a flywheel, which stores rotational energy and extends the time for riding through the disturbance.

Synchronous motor. An alternative is use of a synchronous motor as the prime mover. This avoids the slip problem. Now, there is a motor rotating at the same synchronous speed as the generator, and the input-to-output frequency is exactly maintained (0% error) since both input and output are locked at the same speed of rotation.

Another valuable attribute of a synchronous machine is its ability to continue running on just two of three phases. This characteristic is very similar to the capability of three single-phase transformers connected in delta: This transformer system will continue to provide 3-phase power with one of the transformers inoperative or failed. This is an open delta connection, and the transformer bank must be derated by 43% to avoid overloading. Nevertheless, you can continue to have 3-phase service with the one failed unit. The synchronous motor is the rotating equivalent of this circuit,

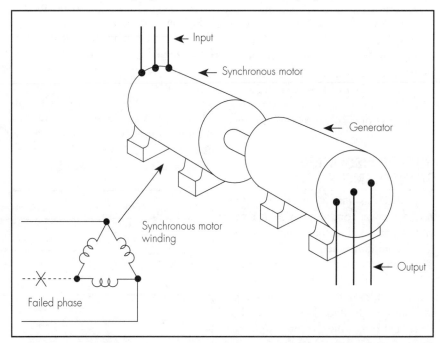

Fig. 8.8. The synchronous motor is the rotating equivalent of a transformer bank of three single-phase transformers providing 3-phase power. The motor continues to operate on the loss of a line connection.

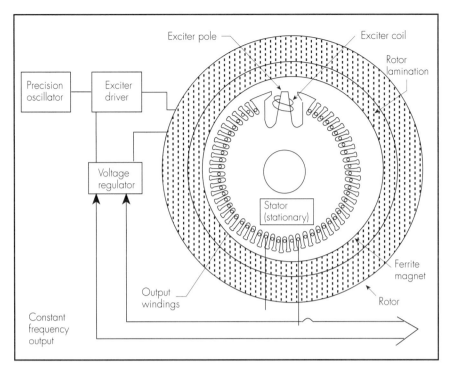

Fig. 8.9. Schematic diagram of "pole writing." Despite variation of rotor speed, the generator's exciter coils continuously print proper poles or AC frequency disposition on a magentized rotor. This, in turn, ensures constant frequency/voltage for noninterrupted power.

but continues to operate not on the loss of a coil, but on the loss of a line connection, as shown in **Fig. 8.8**.

This capability is of value not only when a utility single phasing disturbance occurs, but also during lightning storms. First Phase A goes, then Phase B, then Phase C, and then back and forth during the storm. The capability of running on two of three phases allows the synchronous machine to take power from any two phases at any given instant, thus permitting a ride-through of storm disruptions with no time limitations.

The one flaw of a synchronous motor driven MG set is the lack of starting torque and/or recovery capability of a pure synchronous motor. Its start is assisted by a pony motor, which brings the unit up to synchronous speed. When power is removed from the input, the synchronous motor breaks out of locked speed at about 100 milliseconds into the disturbance. If the outage is within that time frame, all goes well and the machine continues to run. If the time of outage exceeds 100 milliseconds, the unit then shuts down and has to be restarted.

Synduction motor. One development, the synduction motor, goes a long way toward making the MG set a power-conditioning-capable product. This technology consists of updating the prime mover so it is an induction and synchronous motor all in one. This motor has the torque capability of an induction device, remains in the induction mode until almost at the synchronous speed, and then internally switches windings to become a synchronous-running device that is locked at the synchronous speed, thus assuring constant frequency.

Written-pole motor. A product improvement came through the efforts of John Roesel who developed a patented "written pole" technology. This brought a large ride-through improvement by providing a constant frequency output, even while the motor is slowing down when incoming energy from the power line is disrupted.

Normally, as the power is removed from the input to a motor, it will lose speed and sacrifice stable frequency. In this breakthrough, a way has been found to electronically reposition the poles of the alternator to compensate for the slowing speed of the motor. The set can produce full rated output for 15 to 20 sec, without battery support! Sounds too good to be true, but the technology has been working for many years. An example of the pole writing is seen in **Fig. 8.9**. In a two-pole, 3600-rpm application, the poles of the alternator are successively repositioned backward to provide a constant 60-Hz output as the input loses rotation. Thus, making a constant frequency, variable pole machine.

In addition to extending the ride-through time, the Roesel development provides a new combination for continuous power: coupling of an extended-time motor-alternator with a conventional engine generator. In this application, shown in block diagram in **Fig. 8.10**, an

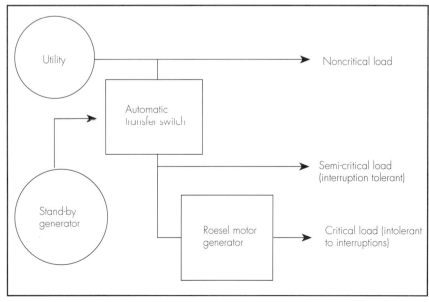

Fig. 8.10. Roesel development uses the combination of an extended-time motor alternator with a conventional engine generator.

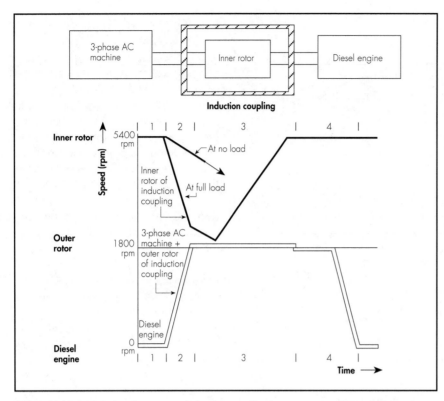

Fig. 8.11. Hybrid design features an induction coupling between an engine and its generator.

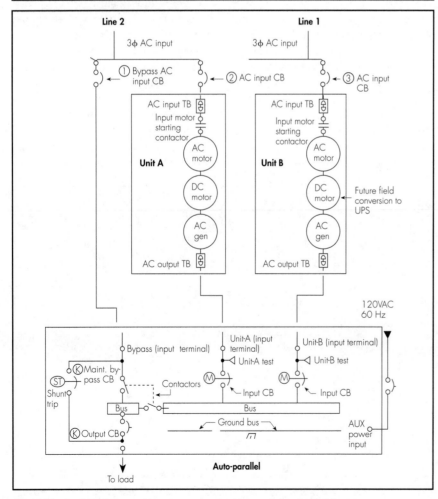

Fig. 8.12. Dual-feed motor alternators connected in parallel on load side.

engine generator and automatic transfer switch (ATS) are located as the alternate source to the utility feed, upstream of the Roesel unit. The ride-through of 15 to 20 sec provides sufficient time to start, synchronize, and transfer online the engine-driven unit without any disturbance to the solid-state electronic load.

Other MG Technologies. A further extension of the coupling of technologies is seen in **Fig. 8.11**. Here, an induction coupling has been inserted between an engine and its generator. This hybrid design is intended to provide continuous power, first as the electric utility energy is used to drive the output, and then as the stored rotational energy is used to start the engine to continue driving the assembly.

In a combination of perhaps the best of both worlds, dual feed motor alternators are connected in parallel on the load side in a fashion shown in **Fig. 8.12**. Here, where they are available, two continuous utility feeds are used to eliminate the need for battery energy in providing uninterruptible power. The layout begins with the use of the synduction technology, in its simplest form acting only as a motor driven alternator on each of two units, equally sized, and connected to separate incoming power feeds. Since the units are synchronous, they can be paralleled without costly synchronizing equipment and can handle the 50% load step that occurs if one line should experience an interruption. The units are constructed with a shaft extension for future field mounting of a DC motor in the event of the need for conversion to self-contained battery-supported units.

The idea of this is to provide only the level of performance needed in the initial installation, but have the full flexibility available to convert a unit in the field without loss of the initial investment. In two large telephone support installations, this conversion was a saving feature for the client. The total cost of adding the conversion is greater than the cost to purchase the finished assembly all at one time, but permits flexibility of application and the opportunity to operate at the lesser level as long as possible.

A magnetic synthesizer uses pulse transformers, inductors, and capacitors to generate a 12-step, 60 Hz output independent and isolated from the power input.

MAGNETIC SYNTHESIZERS

Power synthesis equipment utilizes the incoming power merely as an energy source, from which it creates a new, completely isolated power output waveform to supply to the computer. This synthesized or regenerated waveform is engineered to fall entirely within a computer's tolerance requirements, with an adequate margin for safety, regardless of the disturbances to the incoming power source.

Magnetic synthesizers use ferromagnetics (pulsed transformers and inductors) and capacitors to synthesize a high-quality 3-phase output, using the input only as a source of energy. The output is fully isolated from imperfections in the input power. A 3-phase output of reduced capacity can be maintained if one input phase is lost, although the angle between phases will not be 120°.

Protection. A magnetic synthesizer has no moving parts or semiconductor devices, except in the controls. Enough energy is stored in the capacitors to ride through an outage of about one cycle. When a reliable synchronized alternate power source (such as a second utility line) is available, the magnetic synthesizer has been combined with a static transfer switch that can transfer to the alternate source in ¼ cycle. This provides uninterrupted power to the electronic load and eliminates the need for a UPS.

This equipment can protect against oscillatory transients, sags and swells, overvoltage and undervoltage situations, and voltage surges (by using an arrester accessory). Magnetic synthesizers do not protect a critical load against a complete power outage but, via stored energy, have a ride-through capability of up to 1 cycle. Battery backup is possible if additional operating time is required.

Characteristics. The major characteristics of magnetic synthesizers are as follows.

• Can operate with an input voltage range as low as +/-40% or more of the nominal voltage, which protects against deep voltage sags caused by power system faults.

• Input power factor remains in the range of 0.96 or higher from half to full load.

• Because it regenerates output voltage waveform, output voltage distortion (less than 4%) is independent of any input voltage distortion, including notching.

• Efficiency at full load is typically in the range of 89 to 93%.

• Minimum maintenance is required beyond an annual replacement of failed capacitors.

• Redundant capacitors built into the units allow several capacitors to fail between inspections without noticeable effect on performance.

• Output voltage varies about 1.2% for every 1% change in supply frequency. When supplied by a limited-capacity onsite generator, a 2 Hz change in generator frequency (quite large), causes output voltage change of only 4%, which is of little consequence for most loads.

• Accepts 100% single-phase switched-mode power supply load without derating, including all neutral components. Also isolates sensitive load from power source voltage distortions, and power source from harmonic currents associated with nonlinear loads.

• Input current distortion remains less than 8% THD even when supplying nonlinear loads with over 100% current THD.

• Need for bypass circuit minimized since contains only dry-type transformers and self-healing AC capacitors, with no moving parts to wear out nor electronics to fail.

STATIC UPS SYSTEMS

An alternative to the items discussed thusfar is the solid-state UPS. It provides a nonmoving-part answer in ultimate power conditioning. This technology uses power electronics to convert AC energy into DC, store it in a battery for long term, and then send the DC power into an inverter, changing the DC back to AC. **Fig. 8.13** shows a basic block diagram of these elements, along with a high-speed static switch on the output of the entire assembly.

In critical situations, a static uninterruptible power supply (UPS) can be cost effective. It not only provides line-side filtering and conditioning necessary to assure the power quality required by the solid-state electronic equipment it feeds, but also supplies power for a specific length of time after the loss of utility power. This represents a significant increase in protection for critical loads.

Rectifiers convert raw input AC power to DC, which keeps the floating battery fully charged and supplies power to the inverter section. Rectifiers may use diodes, thyristors, or transistorized high-frequency switching

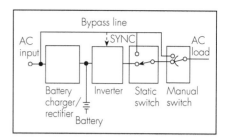

Fig. 8.13. *Basic block diagram shows power electronics that convert AC energy into DC, a battery plant for long-term storage of this energy, and an inverter that receives the DC power from the batteries and changes it back to AC. Also included is a high-speed switch on the output of the entire assembly.*

power supply to provide the DC.

Inverters use thyristors to generate six- or twelve-step waves, pulse-width modulated waves, or a combination of two to synthesize the output. Switching transistors are used in some inverters for high-frequency switching that requires less output filtering. Inverter output should be a stable, low-distortion sine wave.

Batteries supply power to the inverter when input power has been lost or during a period of very low input voltage. Batteries can be sized to provide power for any specific length of time at full load. Usually, this time ranges between 5 minutes and 1 hour, with 15 minutes being the most common. Batteries for large installations are usually lead-calcium wet cells. Some smaller units use gelled-electrolyte cells or, more recently, immobilized-electrolyte cells. Both of these types are sealed, "maintenance-free", and spillproof. Since batteries are connected at all times, there is no switching, and full, undisturbed power to the sensitive loads continues following the loss of input AC power.

Static bypass switches transfer power from the normal source to a backup source, and then from the backup source to the normal source. Each transfer normally takes less than 4 milliseconds, so there is no break in the continuity of power. The choice of what constitutes the normal and what is the backup source depends upon source availability at the jobsite and the configuration of the UPS.

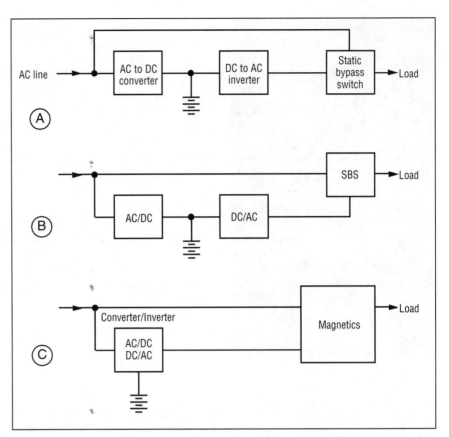

Fig. 8.14. *Block diagram of the three basic types of solid-state UPS systems. (**A**) is a reverse transfer type feeding the sensitive load through the converter/inverter under normal conditions. (**B**) is a forward transfer type that is essentially the same as the reverse-transfer type but normally feeds the load directly from the AC line. (**C**) is the line-interactive type of UPS.*

Maintenance bypass switches are sometimes incorporated into the UPS system. The static bypass switch transfers the computer load from the UPS system to utility power upon failure of a UPS module, which will permit maintenance on the failed module. However, if the static switch fails, or requires maintenance while not energized, a maintenance bypass switch will be essential for providing an uninterruptible source of power. Therefore, this type of switch is usually recommended.

STATIC UPS CONFIGURATIONS

Static UPS systems are available in three common configurations: on-line; off-line, and line interactive.

UPS systems *come in a wide range of ratings, from VA-sized units that are applied between the VDT and the computer (left), to large 750kVA lineups that protect power to mainframe computers (right).*

On-line UPSs (also called "reverse-transfer") shown in **Fig. 8.14(A)** are the types most often used for larger installations of critical loads. In this configuration the normal supply of power feeds the rectifier, and the load is continually supplied by the output of the inverter. Upon loss of the inverter output, or its failure to produce an output that is within specifications, the load is switched to a backup source, usually the utility's line.

This alternate power source serves a second purpose: it provides a "stiff" current source for clearing faults, an important function because most inverters cannot provide the large amount of current required to blow a fuse or trip a circuit breaker on a branch circuit feeding a critical load. The static bypass switch retransfers the load back to the UPS output when the protective device clears the fault.

The advantage of an on-line UPS is that the critical load continually receives conditioned power and never sees a loss in power when the prime source is lost. The inverter will simply draw its power from the battery bank. One important requirement is that the rectifier in an on-line UPS must be adequately sized to meet the input requirements of the loaded inverter as well as maintaining the battery's charge.

There are four steps to the operation of an on-line UPS, as shown in **Fig. 8.15**. In normal operation (**A**), the rectifier-charger provides DC current to the inverter while maintaining the batteries at full charge.

The emergency phase (**B**) begins when the utility power fails. As the AC line voltage drops, the output of the rectifier-charger also drops. The batteries compensate for this decrease and, as a result, the inverter output remains unchanged. The drop in AC input to the rectifier causes the batteries to reverse-bias the rectifier diodes thus preventing battery current from flowing back through the charger. The batteries continue to drive the inverter until AC power is restored or the battery output drops below the level required to drive the inverter. This depends on the battery ampere-hour rating and the size of the load.

The bypass phase (**C**) occurs when the UPS malfunctions, or when the inverter's output does not meet specifications. The static bypass switch automatically transfers the load to the alternate source (usually the utility line) within 4 milliseconds.

Because most static bypass switches are not rated to continuously carry the full load, a circuit breaker is usually connected in parallel with the static bypass switch as shown in (**D**). Being

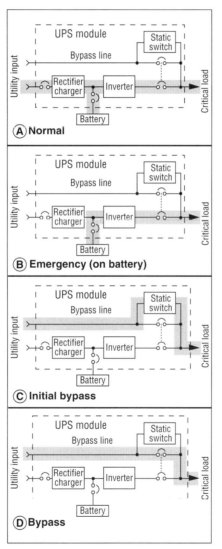

Fig. 8.15. Four stages of a reverse-transfer UPS operation. **(A)** is the normal flow of power to the critical load. Upon failure of AC power or rectifier failure **(B)**, the battery supplies power to the inverter. If the inverter fails to produce the proper waveshape, the static transfer switch interrupts power from the inverter and switches to the AC line **(C)**. Because the static transfer switch is not usually rated for full load, a parallel electromechanical circuit breaker also closes **(D)**.

an electromechanical device, the circuit breaker is very much slower in closing. When it does close, however, it has adequate current carrying capacity to accommodate the critical loads served by the UPS.

A return of normal-source power is sensed by the UPS's control circuitry and a retransfer to UPS output takes place automatically.

Off-line UPSs (also called "forward-transfer") shown in **Fig. 8.14(B)** are different in that they allow raw unconditioned power (unless conditioned by upstream equipment) to feed the solid-state electronic loads. The static transfer switch only directs the inverter AC output to the load when it senses a fluctuation or interruption of the normal source. This configuration is favored for smaller power applications and where power conditioning is not normally needed. It should be remembered, however, that the transfer time between normal and alternate sources takes place via the static bypass switch in 4 milliseconds. For equipment adhering to the ITIC (CBEMA) curve, this loss of power should be tolerated.

Line-interactive UPSs are shown in **Fig. 8.14(C)**. Line power is not converted into DC, but is fed directly to the critical load through an inductor or transformer. Regulation and continuous power to the critical load is achieved through the use of inverter switching elements in combination with inverter magnetic components such as inductors, linear transformers, or ferroresonant transformers. The term "line-interactive UPS" comes from the fact that the inverter interacts with the line to buck, boost, or replace incoming power as needed to maintain a constant uninterrupted voltage to the sensitive load.

OTHER STATIC UPS CONSIDERATIONS

Input power factor. The uncorrected input power factor of static UPS systems is typically 80%. This value can be increased with the addition of a capacitor circuit to any desired value (93% to 95% for full-load would be desirable). Low input power factor systems require high input current compared to high power factor systems. Higher currents require increased ratings of input overcurrent protection devices and cable sizes, and might also limit the capacity of the total UPS system. The drawback in adding capacitors is that it places additional devices on the electrical system along with their inherent potential to fail.

Harmonic currents. Static UPS systems will produce undesirable harmonic current on both the input and output lines because of the firing circuits of the silicon controller rectifiers (SCRs) in generating the desired waveform. Total input and output harmonic distortion is typically 12% and 5% respectively. If the power system serving the UPS system equipment does not contain harmonic dampening equipment (motors, transformers, etc.), the input harmonic content may effect sensitive equipment not on the UPS system, and will adversely affect the governor control system and overheat the neutral conductor on local emergency generating equipment. Therefore, input filtering may be required to limit the input harmonics.

Don't forget the equipment's ability to handle current harmonics on the load side of the UPS without making voltage distortion disturbance across the entire output bus.

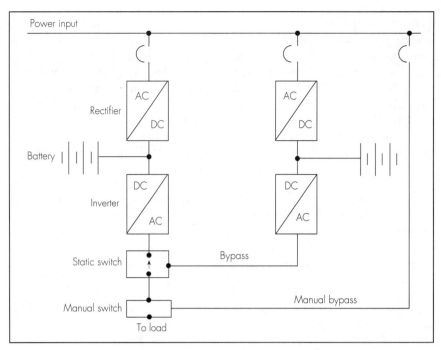

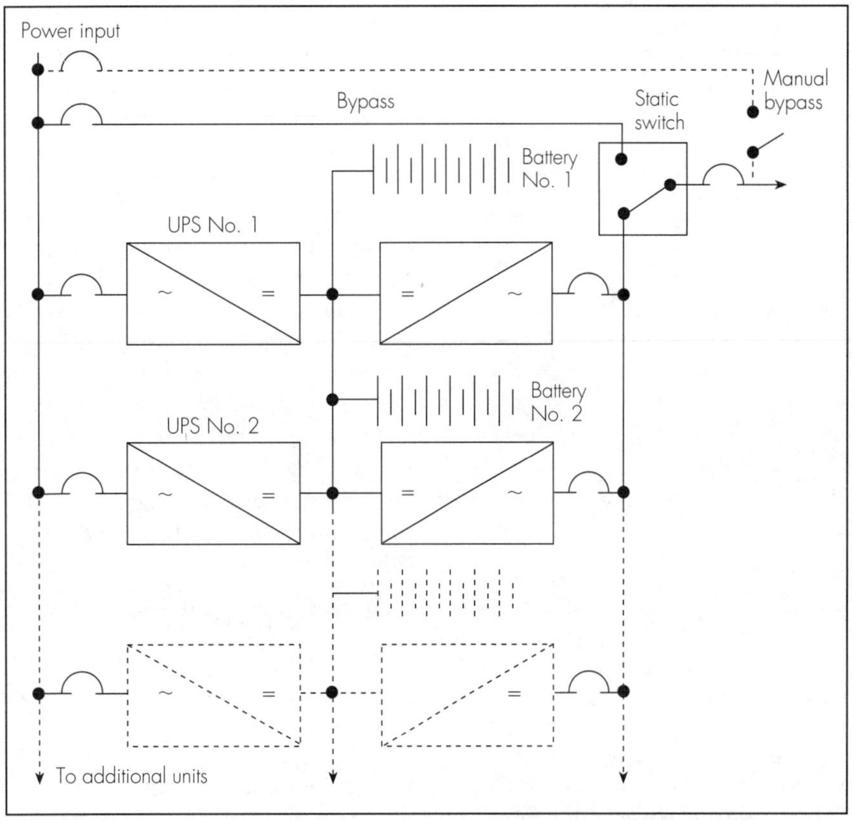

Fig. 8.17. A second complete UPS unit is installed in the bypass line so should the first system require transfer to bypass, the bypass itself would be a conditioned supply with battery backup.

STATIC UPS REDUNDANCY

Since the prime concern for UPS technology is uninterruptibility, several techniques have been applied to make the equipment more reliable if any single part or element fail. Designs that incorporate redundancy are well worth considering, especially for very sensitive loads or operations where downtime is extremely costly or not tolerable at all.

In **Fig. 8.16** there are several elements operating in parallel in order to compensate for any single failure. Notice there is no need in normal operations to transfer to the unconditioned bypass line, since the additional elements in the system keep the load running by themselves. The bypass is available for catastrophic failure and with a 4-ms static switch just as before.

The key in this arrangement is to have one more rectifier/inverter package than is needed to supply the load rating. For example, the system could include three sets of rectifiers, inverters, and batteries, with each set to be

Fig. 8.16. In this extension of Fig. 8.13, there is no need in normal operation to transfer to the unconditioned bypass line, since the additional elements in the system are to keep the load running by themselves. The bypass is available for catastrophic failure and is supplied by installation of a 4-ms static switch.

operated in parallel. Any two of these sets could be counted on to furnish rated power to the load at any time. All the sets would be running and sharing the load, but no harm would come to the system if one set failed or required service. It could be disconnected from the output bus using an equally fast method of static switching known as a *static interrupter*. No disturbance to the system would be seen as that part of the UPS was serviced and restored to the system. This sequence could continue until all three sets were serviced or examined, and the load would still be powered by fully on-line UPS.

Another form of redundancy is seen in **Fig. 8.17**. Here, a second complete UPS unit is installed in the bypass line. Should the first system require transfer to bypass, the bypass itself would be a conditioned supply with battery backup.

When budget restraints limit the extent of redundancy, be sure the UPS has at least two strings of batteries. The weakest link in the system is the battery, and having the battery power in two strings of cells gives protection against a single cell failure that will disable the UPS entirely.

Another important consideration is the equipment's ability to handle current harmonics on the load side of the UPS without causing voltage distortion disturbance across the entire output bus. Also, to avoid interaction problems with the facility power supply, be sure to find out what the harmonic current distortion is on the incoming line to the rectifier.

ROTARY UPS SYSTEMS

A rotary UPS system (**Fig. 8.18**) is basically an MG set that has been adapted to providing AC voltage to solid-state devices even when prime source power is interrupted.

Rotary UPS systems vary more in concept and design than the static UPS. One thing most rotary UPS systems have in common is that the output to the load is supplied by an MG set during normal operation and on loss of normal power. The principal advantage of these systems is that the output of an MG set is a clean, low-distortion sine wave, completely isolated from disturbances in the input power supply to the motor.

On loss of normal power, an MG set (sometimes with a flywheel) has enough inertia to continue supplying the load for a period of a few cycles to several seconds. During "ride through," an auxiliary source of power must be brought on line to supply the motor that drives the generator in order to maintain uninterrupted power to the loads. In some systems this auxiliary source is a battery set that floats on the line at

This small low-noise-producing rotary UPS system is installed directly in a computer room.

all times. This auxiliary battery power can be supplemented with an engine-generator brought on line before the battery power is exhausted. In some cases, the system ride-through is long enough to permit the engine-generator to be brought on line directly (usually 10 seconds or more), and intermediate power from batteries is not required.

Some systems have backup power supplied by a rectifier-battery-inverter system similar to a static UPS, but with the inverter output driving the AC motor of the MG set. Since the generator puts out clean power even with poor-quality power input to the motor, the rectifier-inverter can be a simple, low-cost unit with little output conditioning.

Some available rotary UPS systems include:

• A single-shaft synchronous MG, with a common rotor and stator winding, supplied by a rectifier-inverter system and battery backup. In normal operation, 95% of the motor power is direct from the utility source and only 5% from the inverter operating in parallel with the utility. On loss of utility power, the motor is supplied 100% from the on-line inverter, powered by the battery, with no transfer time required.

• DC-motor-driven MG, with the DC for the motor supplied from a solid-state rectifier, which also keeps a floating battery charged. On loss of normal power, the battery powers the motor di-

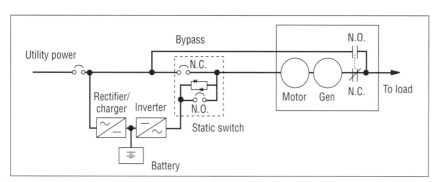

Fig. 8.18. *Block diagram of a typical rotary UPS system.*

Hybrid rotary UPS equipment provides the ability to ride through extended power outages.

rectly, with no transfer time. This is a UPS with the MG set acting as a rotary inverter.

• 3600-rpm induction-motor with a flywheel, driving an 1800-rpm generator through belts and adjustable-speed pulleys. The ride-through on this system is more than 12 seconds, allowing an engine-generator to be brought on line without intermediate battery power. Other units with 1800-rpm induction motors and adjustable-speed belt drives use batteries and inverters to drive the motor on loss of normal power.

• Synchronous MG, with an additional DC machine on the same shaft. With normal power, the DC machine coasts, with controls driving it as a low-output generator to keep a backup battery charged. On loss of normal power, the control system switches so that the battery powers the DC machine as a motor, driving the generator while the AC motor coasts.

• A vertically-mounted MG, with a large flywheel providing enough ride-through to bring an engine-generator on line. The engine-generator is off until normal power is lost; then is started and on line within two seconds.

• A variable-speed, constant-frequency synchronous MG. The machine varies the number of poles of both motor and generator, so that the frequency remains the same over a wide speed range. On loss of normal power, generator voltage and frequency output at full load remain within computer limits for 25 seconds as the machine slows down, providing sufficient time to bring an ordinary engine-generator on line. No batteries or inverters are required.

HYBRID ROTARY UPS

Another form of rotary UPS is a hybrid system, such as the one shown in **Fig. 8.19** that connects the load to the normal supply through a large inductor. Tapped to the inductor is a synchronous motor, overexcited in normal operation to act as a rotary condenser, which with the inductor acts as a resonant filter to provide clean power to the load. In normal operation, the synchronous motor rotates a magnetic clutch. On loss of normal power, the magnetic clutch provides enough inertia for the synchronous motor to act as a generator and supply the load. A diesel engine is brought up to speed in less than one second and connected by a separate mechanical clutch to drive the synchronous motor.

When utility power fails, power to the AC windings of the induction coupling is cut off while the DC winding is being excited, which causes the inner rotor to decelerate and transfer its stored kinetic energy to the outer rotor. As a result, the synchronous machine's speed is kept at 1800 rpm and its function now changes from a motor to a generator. While all of this is going on, the diesel engines starts under no load and accelerates to 1900 rpm in approximately 1.5 seconds.

As the speed of the inner rotor matches the speed of the diesel engine, the free wheeling clutch engages the engine, which now becomes the driving energy source. In order to maintain a constant load frequency of 60Hz, the DC winding excitation is precisely controlled. This, in turn, regulates the torque transmission between the diesel engine and the synchronous motor.

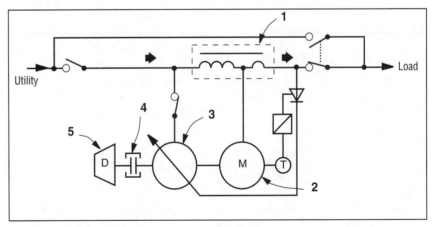

Fig. 8.19. Another form of rotary UPS combines the functions of a UPS and an emergency generator. Its major components include: (1) an autoforming inductor (choke); a synchronous motor (2); an induction coupling (3); a diesel engine (5) with a free-wheeling clutch (4). The synchronous motor is fitted with an induction coupling that interfaces with the diesel engine. This coupling has an outer rotor that is mechanically coupled to the synchronous motor and fitted with a set of 3-phase AC windings and a DC winding. The inner rotor is separated from the diesel engine by the free-wheeling clutch.

SITE ANALYSIS AND MONITORING

Power systems problems that affect sensitive electronic equipment must be corrected in order for these loads to perform well. But before a solution can be implemented, it is first necessary to identify the source of the predicament. With the increasing use of equipment that represents a nonlinear load, the difficulty is often harmonics related. In this chapter, various types of approaches and equipment that are available for measuring, recording, analyzing, and identifying the cause and location of the problem will be discussed.

Accurate electronic measurement of rms values has been made practical by microprocessors. RMS measuring circuits sample the input signal at a high rate of speed, typically about 100 times the highest harmonic frequency. To measure the 25th harmonic of a power system (a frequency of 1500 Hz) the sampling rate would be about 150,000 times per sec. The microprocessor circuits digitize and square each sample, add it to previous samples squared, and take the square root of the total. This is an accurate rms value, regardless of the waveform being measured.

DETERMINING POWER QUALITY

In most instances, diagnosing and solving disturbance problems at the end-user's facility can be accomplished through a systems approach. This involves the evaluation of specific areas such as wiring, grounding, and surge protection. Electrical wiring, grounding errors, or loads in other areas of a

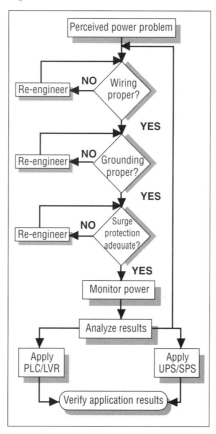

Fig. 9.1. A systems-approach flow chart for determining user needs.

facility generally account for the majority of disturbance problems. Remember, especially with regard to grounding and surge protection, that it is how these protective systems are applied back to the service entrance of the building that is important. It is possible the distribution system supplying the solid-state electronic equipment is properly installed, but the basic infrastructure supporting this equipment may be flawed. For more details on what to look for, see Chapters 6 and 7 in this book, as well as EC&M's companion book to this one entitled, *Practical Guide to Power Distribution Systems for Information Technology Equipment*.

If the problem is not solved at the primary power distribution level, then the input power to the solid-state electronics should be monitored for a period of time, usually a normal business cycle. Here, a recording power and/or harmonic analyzer is used. By analyzing the monitor's graphic data, the cause and effect relationship can be sorted out. Based on this analysis, an intelligent decision on the proper conditioning equipment required can be made. The flow chart in **Fig. 9.1** illustrates this troubleshooting method.

INITIAL STEPS

The very first steps in conducting a

site analysis involve gaining familiarity with the site, its power distribution, and its loads. While relatively simple, these steps are extremely important.

Visual inspection. Walk through the site, looking closely at all electrical systems and distribution equipment. Take notes and, if permitted, use a video recorder or camera. Note the load equipment and its relationship to the building wiring system supporting it. For example, maybe a line of variable-speed drives and PCs in a room are powered by dropouts from plug-in busway that serves both loads.

Don't pass up the chance to look inside equipment. In fact, the fronts of panelboards should be taken off, and doors of cabinets, MCCs, etc. should be opened to see what their interior wiring looks like.

Single-line diagram. Reviewing this document is very important because it gives a better picture of power distribution at the site. The problem, however, is to make sure these drawings accurately represent the present system. At a minimum, spot check several items to confirm these are "as-built" drawings.

Connection points. Use the single-line diagram to help identify electrical system parts to which test equipment is to be connected. For future reference, make sure tested equipment points are identified in test reports by the same names or numbering as shown on the single-line diagram.

Power sources. Make sure to identify each AC power source serving loads suspected of being affected by power-quality problems, using the one-line diagram as the reference. Find out if the power source serves a single load or multiple loads. If multiple, the operation of the attached equipment not under investigation must not affect or mask the measurements being taken.

INITIAL TESTING FOR HARMONICS

After completing the preliminary steps, it is time to start taking measurements. Most importantly, make sure the test equipment is placed where it is away from locations where it can be damaged, and test leads are connected in the most electrically safe manner. This is especially important when leaving the test equipment on-line, unattended, and with covers of equipment removed or open. Temporarily covering the equipment with rubber blankets or similar nonconductive protection is an even better idea. And don't forget to wear safety glasses.

The first set of measurements can usually be made with handheld instruments since most power problems normally can be identified using them. Problems that are highly intermittent or involve simultaneous events on multiple conductors should be done after these preliminary measurements are taken. The following items should be checked.

Line voltage. With the load deenergized, take a voltage reading at power-source outputs and at their respective load input terminals. Next, energize the load and take the same readings, Then calculate the respective circuit's voltage regulation using the following equation:

% reg = $100 \times (E_{loaded} \div E_{unloaded})$

The NEC sets a 3% maximum limit on the percent voltage drop from the origination of a branch circuit to the furthest outlet. To see what's happening on the tested circuit, calculate the voltage drop to the specific load under test by using the following equation:

$\%E_{drop} = (E_{unloaded} - E_{loaded}) \div E_{unloaded}$

Verifying a circuit's operating conditions, however, will not give the overall picture for the circuit. What is still unknown is what's happening to the circuit during starting or inrush-current conditions. This can be determined by using a recording voltmeter (often available as a feature of a handheld voltmeter). Capture the conditions during the first few moments of the load being energized. Excessive voltage drop or poor voltage regulation typically means there is too much impedance in the current path.

Typically, fixed high impedances in a circuit include the internal impedance of transformers, wiring connections along the wiring path, and the conductors themselves being too long, etc.

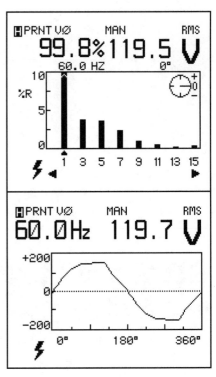

Fig. 9.2. Screen display shows the nearly sinusoidal voltage waveform of a nonlinear load, along with a bargraph of the harmonic spectrum to the 15th harmonic.

If an active source of voltage is part of the circuit (an engine-generator set, MG set, or UPS), the source's own internal impedance and voltage regulator's characteristics may be the problem, not the downstream paths with their fixed impedances.

If the meter being used has waveform capture-and-display as a feature, obtain the voltage waveform, voltage regulation and voltage drop conditions. **Fig. 9.2** shows a voltage waveform and a bargraph of the harmonic spectrum to the 15th harmonic for a specific nonlinear load.

Line current. Measure starting inrush and normal operating currents of the load under test. These should be made using an appropriately ranged clamp-on ammeter or recording multimeter. If the meter being used includes waveform capture-and-display provisions, get a graphic of the current waveform and harmonic spectrum. **Fig. 9.3** shows the ones generated for the same nonlinear load as Fig. 9.2.

With graphics of both voltage and current conditions, compare the two wave forms and look for any interac-

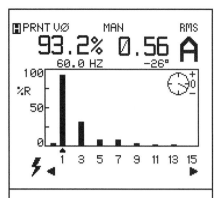

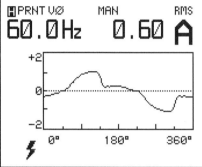

Fig. 9.3. *Screen display shows the nonsinusoidal current waveform of the same nonlinear load, along with a bargraph of the harmonic spectrum to the 15th harmonic.*

tive effects of the load with the AC supply circuit impedances.

Neutral-to-ground voltage. There's a difference of opinion on the importance of making neutral-to-ground or neutral-to-chassis voltage measurements, especially where branch circuits connect to load equipment. Some contend that a measurement of a few volts indicates a problem in need of correction.

What actually is happening when such a voltage is registered is the meter is reading the algebraic sum of two possible voltages. One is the neutral wire's voltage drop due to the load current; the other is the voltage difference between two points (where the AC power source's system neutral is grounded and where the local branch circuit equipment ground is installed).

With the load equipment *deenergized*, make a voltage test between the neutral and ground or chassis. This chassis measurement is sometimes useful because it can show the presence of common-mode voltages between the AC system's ground and the local ground where the load equipment is installed. These common-mode voltages may be a symptom of an electrical noise problem on associated data, signaling, or communications cables. Unless these voltages cause related common-mode currents to flow on the equipment's signal-level circuits, they won't have any effect on the operation of the load equipment itself.

With the load equipment *energized*, the voltage test picks up the voltage drop in the neutral wire due to the load's operation. On a single-phase, 2-wire-plus-ground circuit, this is the same as for the hot or line wire. This voltage, however, is of no consequence to the equipment's operation, except as it relates to poor voltage regulation on the circuit.

ADVANCED TESTS

Some advanced testing is needed to see if harmonic voltages and currents are a problem, if impulses (transients) are present and what they are doing to the load equipment's operation.

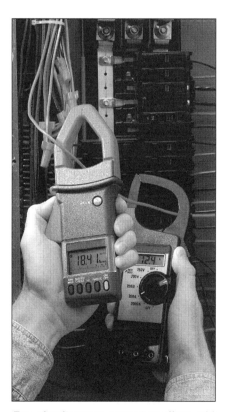

Two simultaneous current readings *with either an average-responding or peak-reading clamp meter and a true-rms clamp meter on the phase conductor feeding the circuit should be compared. If there's a significant difference between the two readings, lower order harmonics are present in the circuit.*

Simultaneous current readings. There's one simple test that can be used to see if lower-order harmonics are present in a single-phase branch circuit. It involves making two simultaneous current readings with either an average-responding or peak-reading clamp-on meter and a true-rms clamp-on meter on the phase conductors feeding the circuit.

If there's a significant difference between the two readings, the presence of lower-order harmonics in the circuit is likely. This two-current comparison method usually works for single-phase circuits because single-phase nonlinear loads typically have current waveforms with sharp peaks, which cause the true-rms value (heating equivalent) to be higher than the average value. This technique, however, cannot be used reliably on 3-phase nonlinear loads because of the typical double-pulse nature of the 3-phase current waveform.

Crest factor. Another relatively simple test is to determine the circuit's crest factor. In a pure sinusoidal wave the ratio of its peak to its rms value (crest factor) is 1.414. The relationships are shown in **Fig. 9.4**.

To determine a circuit's crest factor, use a true-rms meter having instantaneous peak capture capability to measure voltage. A good voltage sine wave will have a peak value that is very close to 1.414 times the true-rms value. So, a 120V rms sine wave will have a peak value of approximately 169.7V. A voltage peak significantly higher than this is an indication of harmonics being present.

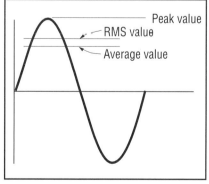

Fig. 9.4. *Values (sine wave only):*
V_{rms} *or* I_{rms} *= 0.707 × peak value*
= 1.11 × average value
Crest factor = peak/rms = 1/0.707 = 1.414

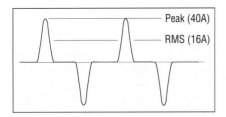

Fig. 9.5. *Values for nonsinusoidal voltage or current rms or peak values are obtained by measurement; values will change with pulse shape, duration, frequency, and peaks. For example, if*
  peak = 40A and rms =16; then
  Crest factor = 40/16 = 2.5

Care, however, must be taken in applying this test. For a pulsed wave of the same peak value, but with considerable "off" time and a low rms value, the crest factor will be much higher (as per the example shown in **Fig. 9.5**).

Voltage waveform peaks. Still another test involves evaluating voltage waveform peaks, since many nonlinear loads will cause these peaks to be reduced or clipped, especially if the source impedance is high. The following is a two-step method to test the voltage for clipping due to harmonics.

• Measure the true-rms value and multiply by 1.414 to find the theoretical peak value; then
• measure the actual peak value.

Compare the actual value to the theoretical value. If the values are significantly different, the waveform is distorted and contains harmonics.

Harmonic disturbance analyzers are especially useful because they provide the ability to view the waveforms, measure values, and record power quality problems.

Three-phase test. For 3-phase, 4-wire wye-connected circuits, attach a CT to the neutral conductor. When phase-to-neutral nonlinear loads are attached on these circuits, there's normally a lot of 3rd harmonic current flowing on the shared neutral. This is where a meter's frequency counter function comes into play.

In addition to measuring the shared neutral's frequency, compare its current to those of the phase conductors. If neutral current is present and the line currents appear to be well balanced, harmonics are present because with linear loads there would be little or no neutral current. With nonlinear loads, this in not true. As a matter of fact, when triplen harmonics are present, they could cause the current to be higher than those of the phase conductors.

WAVEFORM AS A DIAGNOSTIC TOOL

Suppose harmonics are present on the circuit under test. The next step should be finding what they're like. This is where the harmonic analyzer comes into play. With the use of a CT, this instrument can be used to read harmonic currents and voltages. Either a single-channel input meter equipped with memory to store single current and voltage readings, or a multichannel meter can be used.

Harmonic analyzers allow selectively tuning them to the frequency of each harmonic present, usually up to the 32nd order. For each harmonic, its amplitude can be digitally displayed. Some are capable of simultaneously displaying the order and amplitude of several harmonics. Others are equipped with a data port that can be connected to a PC, which in turn enables the data and waveform analysis to be recorded.

To capture high-frequency events or ones that are transient in nature, a waveform-recording instrument is needed capable of being triggered by a power quality event. A digital-memory oscilloscope can be used, but better still is a powerline analyzer.

Handheld oscilloscope equipped with two channels and a combination of voltage and current probes on its input also can find most power quality problems.

One of the best ways to determine where a power quality problem exists relative to the point of testing, is to use the oscilloscope with one channel connected for line-voltage input and the other for a current CT input from the same line. This allows direct comparison of variations in line voltage with the downstream load's input current. If both the line voltage and current increase or decrease at the same time on the same circuit, the conclusion is the power quality problem is upstream from the test point. If the line current increases and the line voltage decreases, the problem is downstream from the test point and involves the load.

A trigger point setting capability is particularly useful. The voltage trigger points can be set to ignore normal events that do not exceed the set point. When a higher voltage is input, the instrument will trigger and record the voltage trace, capturing the event in its memory. With this capability, additive polarity impulses or swells on the AC line can be easily detected. The oscilloscope's input filters can also be set to exclude DC, low, or high frequencies.

COMPUTER ANALYSIS

Once a system's characteristics have been recorded, it is necessary to analyze the results and decide on what corrective actions must be taken. This is often done by sorting through the data captured during the monitoring phase of the project. Often the skill within the facility does not exist to set up the test program and later to analyze the results. Fortunately, computer software is available to act as a guide to establishing the specific program needed at a particular jobsite, and for analyzing results.

Computer programs allow drawing from a database of case histories and industry experience. Alternatives for addressing the needs of the sensitive electronic equipment in question can be examined.

Typical computer software available on disks for this purpose include a provision for conducting a "power quality audit." The audit procedure examines each portion of the survey in detail for each of the following categories:

• initial problem and site visit;

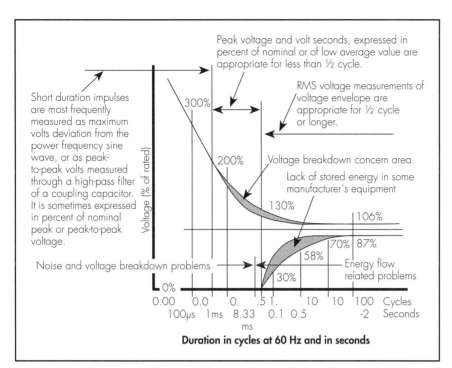

Fig. 9.6. A decision can be made on the type of power conditioning equipment needed to solve a power quality problem by placing event dots onto the CBEMA curve and looking at the dots relative to the curve's boundaries.

- wiring and grounding;
- voltage, current, and impedance monitoring;
- surge/transient analysis;
- harmonics interaction survey;
- power monitoring analysis; and
- power conditioning recommendations.

In providing this in-depth approach to an analysis, the software draws upon all of the national standards available in the industry (ANSI, FIPS Pub 94, NEC, etc.). Each standard that is updated or added to is automatically included in a software update program.

A shorter, "troubleshooting" menu asks for the following review sequence:
- symptoms of the problem;
- evidence of the disturbance;
- voltage, current, and impedance values; and
- power conditioning recommendations.

This shortened audit uses "expert reasoning" to direct the investigator to search for the specific problem based upon the "symptoms" and "evidence" provided by the answers given to questions asked as part of the program.

CBEMA (ITIC) CURVE ANALYSIS

Although the curve shown in **Fig. 9.6** is not a requirement for original equipment manufacturers or end users of equipment per any standard, it can be included as part of a performance standard by contract. In general, most reputable manufacturers build equipment that meet or exceed the limits set by this curve.

Built-in curves. Some makes of powerline analyzers incorporate the CBEMA curve within their software packages, allowing plots to be generated of power quality events relative to the curve. This capability can prove invaluable when preparing a power quality report.

Manual plotting. If access to such software is not available, it's possible to manually place PQ events onto a CBEMA curve. Plot the events using at least three colors:
- Red for events known to cause problems with equipment;
- green for events known not to cause any problems; and
- yellow for uncertain events.

By examining clusters of event dots, the immunity or susceptibility of load equipment to the respective events can be determined based upon their color. For example, a marginal problem exists if clusters of yellow dots are associated with nearby red ones. In addition, write the number of PQ events next to the dot. In doing so, quantitive data that's just as important as the type of PQ event is provided.

Analyzing results: A decision on the type of power conditioning equipment needed to solve the PQ problem can be made simply by looking at the dots relative to the curve's boundaries. For example, any red dots to the left of the 8.33 ms line denote subcycle events, usually impulse- or transient-voltage types.

Those red dots above the curve add voltage or energy. For these, transient voltage surge suppressor (TVSS) protection is required.

Those red dots below the curve denote events that subtract voltage or energy. Here, capacitors or a LC-based filter is recommended because these devices store energy and can release it to fill in the "notches" where the subtractive event has removed energy from the line.

Those red dots to the right of the 8.33 ms line denote rms events that could involve swells, sags, or long-term high or low rms line voltages. For these events, a line voltage regulating transformer, motor-alternator set, or even a full UPS installation is recommended, based on the specific type and extent of the problem and sensitivity of facility processes.

Red dots anywhere along the bottom 0% voltage baseline denote voltage loss events, either transient or rms in nature, depending on whether they are to the left or right or the 8.33 ms line. Here, only a motor-alternator set or UPS can provide the necessary protection from these kinds of events.

Selecting equipment. The plotted events can be use to find the most cost-effective form of power conditioning equipment needed. The following steps are involved.
- Draw a CBEMA curve on a transparency;
- shade the area on the curve representing the proposed power condition-

ing equipment's guaranteed performance;
- overlay this transparency onto the CBEMA curve that has been event marked; then
- determine how many of the red (and yellow) events are fully covered by the shaded area.

Full coverage of the red and yellow dots mean full protection is offered by the proposed power conditioning equipment. This method can be used to see if partial protection, in lieu of full protection, from most of the red-marked PQ events is worthwhile.

MEASUREMENTS

Presence of harmonics (waveform distortion, particularly from the 3rd harmonic), as discussed previously, can be detected by taking rms and instantaneous peak readings of either current or voltage. Since all harmonics generated by nonlinear loads are current generated, the current reading usually will be more sensitive than voltage readings.

Special note: current readings must be taken at the power source of the nonlinear loads while voltage readings can be taken almost anywhere on the bus.

In measuring current and voltage where harmonics are present, true rms reading instruments must be used. These meters are usually calibrated in rms amps or volts, based on a crest factor of 1.414.

On the other hand, conventional meters (analog and digital) measure either the average value or the peak value of a waveform, and then are calibrated to read the equivalent rms value. The average-value measuring meter is usually termed an average-responding or average-calibrated meter. The peak-value measuring meter is usually termed a peak-sensing meter. These meters will give misleading readings when harmonic distortion is present. For example, on a square wave, the average-calibrated meter will read rms values about 11% high while the peak-calibrated unit will read about 30% low. For pulses, the errors can be tremendous, depending on the height of the peak and the off-time between pulses. The average-calibrated meter will read very low, (as much as 50%) while the peak-calibrated unit will read very high (sometimes more than 100%).

Phase current measurements. If the phase current readings of a true-rms sensing meter are significantly different than those of an average-responding meter, it's likely that harmonics are present and are distorting the current waveform. The difference in readings is a function of how the two types of meters measure, as previously discussed. This method is particularly effective when the 3rd harmonic is present.

Neutral and phase current comparison. Another method to identify the presence of triplen harmonics is to measure the phase and neutral currents with a true rms meter. If the neutral current is greater than what would normally be the unbalance of the phase currents, 3rd or triplen harmonics are present.

Oscilloscopes. Another effective way is to look at the voltage and current waveforms with an oscilloscope. Current transformers (CTs) are required to view the current waveforms. If the rms reading is to be accurate for high-order harmonics and pulsed-currents, the CTs must be of high quality (with a very high band width) to sense the high and low frequencies accurately. This is not a problem for pure 60-Hz sine waves.

Spectrum analyzers. A spectrum analyzer can record the current and voltage waveforms, determine the magnitude and types of harmonics present, and provide a printout of these data. If equipment under evaluation has its power supplied from on-site emergency generators, readings of the current and voltage waveforms at the equipment should be taken with the generators on line. This will probably be the worst case because onsite emergency generators typically represent a greater source impedance than the utility system. As a result, higher voltage distortion on the facility's electrical system can be expected when the emergency generators are running.

PRACTICAL TIPS

There are some practical guidelines involved in conducting a typical site analysis. They are based on frequently found disturbances and their characteristics.

Mystery disturbances. First, examine a site for those upsets to a process or solid-state electronic equipment operation that don't seem to correspond to any identifiable source of power disturbance. Such things as ground loops, high speed transients, lightning, and common-mode electrical noise come to mind. Many of these events are here and gone in such a short time frame that they are not easily identified, except with a power disturbance analyzer using high-speed wave shape or event capture.

Repetitive, cyclical disturbances. Look for disturbances that do have a connection with the power distribution system, both within and outside of a facility. These problems will be repetitive and cyclical in nature, definitely power-related, and line-to-line. Examples include voltage sags and surges, momentary interruptions by circuit breaker operations, and power interruptions. It's important to know which disturbance is being faced.

Harmonic distortion. Look for disturbances related to the integer multiples of the fundamental power frequency (60 Hz), the area called harmonic distortion. This area is a subset of the power related area, since these harmonic currents and voltages are recurring. Remember the basics of 3-phase harmonic producers such as VFDs, UPSs, and single-phase switching devices such as PCs and electronic ballasts. The current spectrum seen and its effect on system voltages are indicators of the dominant harmonic present. Special tactics for searching out the problem and identifying solutions is needed.

IDENTIFYING GROUNDING MYSTERY DISTURBANCES

A site is bothered by frequent interruptions that seem to occur when nothing is happening on the facility's electrical distribution circuits. Typically, the equipment goes down on a blue sky day, when there is not a utility problem for hundreds of miles. It's easier to understand the disturbance if it occurred during an electrical storm, or when high winds might be blowing the power lines together. But in this case, nothing is happening.

Typically, the story goes something like this: When an interruption occurs one morning, the operations manager calls the customer service engineer for the equipment in question, who then comes to the facility to analyze the problem. This engineer checks the hardware, software, and operating system and finds nothing amiss. He or she might even comment that, "there must be an environmental problem, since all the systems are OK."

When faced with such a case, immediately suspect a ground noise problem. In fact, the history of mystery disturbance problems over the past 15 to 20 years shows that improper facility wiring and grounding are at the heart of most of these problems.

Here is a step-by-step approach that can be used to track down the problem.

• Make sure that all equipment connections have the same contact with the earth reference potential, and that they're at equipotential with the power ground.

• If the problem persists, determine where the system has a ground loop (a connection where the signal ground and the power ground are attached to ground potential at two different points). This is done by taking voltage measurements with a power disturbance analyzer between the power system neutral at a panelboard and the case ground of the equipment in question. Remember, the idea is not to look for the ground in the panel; rather look for the point of signal reference for the device suspected of having a common-mode problem.

• Look at devices specified with clean, private, dedicated, special, or other specially named ground. (For many years, telephone switches, computers, central processing units (CPUs), and computer numerical control (CNC) machines had these separate ground connections, all in violation of the NEC. All these points were subject to ground loops. Corruption of the process signal occurs when the impedance between the two points results in a voltage differential, which drives a loop current through the driver and/or receiver circuits in the signal path.

• With an analyzer connected between the neutral and case ground, it's possible to see if there are multiple connection points, or if the points have the same potential. If the wires from two points go to the same connection point, the analyzer will show little surge between them. If, however, the points are at different potentials, a steep wave front signature of a spike will be seen.

• It's also possible to test for this condition with a simple impedance meter. **Fig. 9.7** shows how and why this works. First, shut down the problem system, disconnect the respective device dedicated ground connections at their source(s) and connect them together in open air. The impedance of this ground connection setup should be infinite because there is an open air gap between the open air connection point and the supposed single-point ground. If the meter reads zero impedance, then a hidden connection exists around the back of the equipment, providing another pathway to ground. What was thought to be a ground reference radial, or direct connection, is actually a daisy-chained ground.

LOCATING UTILITY MYSTERY PROBLEMS

There are events on the utility system that drastically effect solid-state electronic equipment; for the most part, however, the power supplies in use are built to protect themselves from all but the most excessive excursions in power. Consider the operating voltage range of most power supplies: possibly from 6% or 7% higher than normal and from 12% to 14% lower than normal. In other words, solid-state electronic device vendors are stating their equipment will handle stabilizing voltage once the operating voltage is within the range listed. This range is normally well within that which a utility delivers. Even in the area of frequency, large utility systems are very stable at 60 Hz.

Probably one of the most frustrating mystery event is the momentary interruption caused by a circuit breaker or recloser going through its operation, thereby protecting the overall system from a short-term electrical fault. Even when the system is restored in the comparatively short time frame (50 to 60 ms), the absence of power still takes its toll on power supplies used in almost all modern control equipment.

Take the case of a process using microprocessor controls to operate a line of extruders, a forming operation, and a heating process. All of a sudden, the utility's circuit breaker goes through an operation and the production line shuts down, most of the time with the disastrous results. Is there a major problem

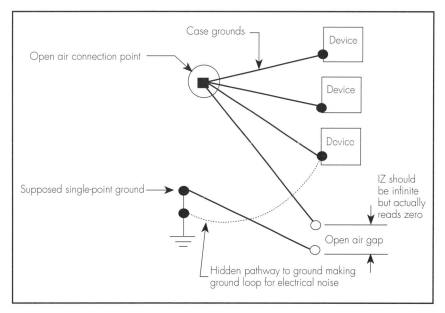

Fig. 9.7. *Impedance of this ground connection setup should be infinite because there is an open air gap between the open air connection point and the supposed single point ground. If the meter reads zero impedance, a hidden connection is providing another pathway to ground.*

with the kilowatts of energy to the heaters, rollers, or forming devices? No. The problem could be the sensitivity of a microchip in a controller section: this chip can't be without power for longer than 8 to 10 ms. At 50 to 60 ms, this chip sees a "loss of power" 6 to 8 times longer in duration than it's capable of withstanding. As a result, it simply informs the machine to shut down.

In beginning a power examination to sort out the many possibilities, it's required to find out what events are occurring, and how the disturbances relate to a power event.

Voltage sagging or swelling. When events correlate with the timing of special equipment functions or large load operation, the need is to verify what range of voltage is actually available on the system. Could the troubled device be getting either too much or too little for some length of time, and thus shutting down to "self-protect?" This is typical behavior of ferroresonant devices when they handle loads beyond their operating range.

When such an event is found, look for voltages that are "out of range." Many times this problem can be corrected with a change of taps on the transformer supplying the equipment. At least, assess how the load operates in the middle of the tap range.

An extended sag is indication of a brownout and will have to be corrected with voltage regulating devices when the condition is beyond the range of transformer tap changes.

Switching transients. Traveling "spikes" or impulses on the power line can cause problems for power supplies, especially those sensitive to sharp rates of change. A common disturbance on a distribution line is the switching of capacitors in or out of the circuit. This action creates a steep, ringing transient that can trick control systems into operation at the wrong time. One result is the tripping of variable-speed drives; the ringing raises the DC bus current level and increases the DC bus voltage above the "trip" threshold, thus causing a false shutdown. In cases such as this, a series reactor or choke would help calm the quick response of the traveling wave and "damp-out" its effect.

Interruption. Complete loss of power for an extended period of time is simple to determine. Many sensitive processes must be able to count on a continuing source of energy. Here the need can be handled by a UPS with an engine-generator back-up. A form of stored energy, either in the battery bank or in the fuel tank or both, helps provide for this continuing power requirement.

When monitoring with a disturbance analyzer to capture the power events, remember that the placement of the monitor is critical to what is being measured. At the service entrance, it's the utility system that is being monitored, but very little of the facility distribution system. As the monitor is moved inward on the local system, the addition is seen of facility effects upon the power conditions. Also, what other loads do on the same bus, or how existing power conditioners behave, can be observed.

LOCATING HARMONIC MYSTERY PROBLEMS

Since the harmonic spectrum is composed of multiple sine waves (three, five, or seven times the 60 HZ frequency), all those requested components will come from the 60 HZ energy source. The power company has a large capacity of sine wave currents, and is able, up to a reasonable limit, to send those different frequencies to the load as it makes its demands.

The concern is that the growing demands of harmonic currents on all electrical systems will result, at some future time, in their incapability to support these needs without some modification of the sine wave shape (distortion for all).

Harmonic currents are "thieves" running on the power systems, robbing electrical capacity. These currents need this capacity in order to run through the system from the supplier to the load

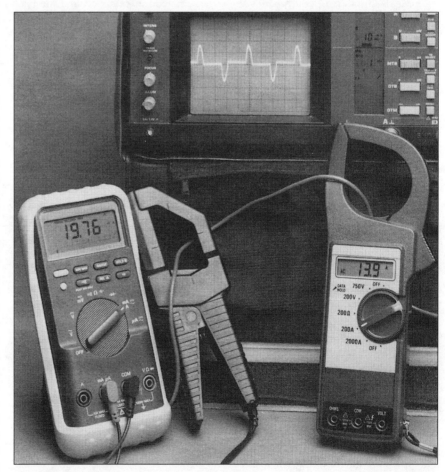

*A **pulsed load** shown on the oscilloscope screen must be measured using a true-rms meter such as the one on the left. An instrument such as that on the right that measures average values and is then calibrated to display the equivalent rms value, will not provide an accurate reading.*

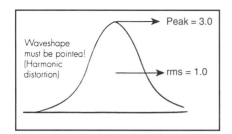

Fig. 9.8. distorted current waveform has a high crest factor; it is outside the suggested maximum 7% range.

device requesting them. They also steal from the power company system if they are of sufficient quantity, and they can cause voltage changes or distortion on entire busway systems as the distorted currents pass through the output impedance of a transformer, making a voltage drop at each harmonic order in the spectrum. The challenge is to keep the thief under control and maximize the work for this size of electrical system. When done, high total PF is achieved.

Some preliminary investigation should be done to see if the problem is present and how bad it is. As was described previously, measure the current at a load device. If it's a 100%, 60 HZ sine wave current, the crest factor (the ratio of its peak to its rms value) is 1.414:1.

Suppose something considerably different is measured, say 3.0:1. This is an indication of a distorted wave shape, since it's so far from the 1.414:1 ratio. In fact, a wave having a crest factor ±7% that of the true sine wave is already a good indicator. As such, set a range somewhere between 1.3:1 and 1.5:1. Reading falling within the limit indicate there is no need to make a further detailed study. If outside this range, get a more sophisticated instrument and make a more detailed study.

Suppose readings shows a high crest factor and the ratio is outside the 7% range, approaching the 3.0:1 peaked wave shape in **Fig. 9.8**. Make preparations to measure the actual current spectrum being asked for by this load device.

This can be done with a hand-held indicating instrument, which will measure the rms value, the percent harmonic current in each harmonic order (3rd, 5th, 7th, etc.), the actual amps at each order, and the total harmonic distortion (THD). Different instruments will do this and more, showing waveshape, capturing the shape, showing bar graphs of the amount of each harmonic, etc.

There are three locations where the extent of harmonic interaction can be determined are: the nonlinear load, the next upstream transformer, and the point of common coupling with the utility.

At the load. Assume it is a variable-frequency drive (VFD). Say the 5th, 7th, 11th, and 13th harmonics, with a spectrum of 40%, 25%, 13% and 8% respectively is measured as the distortion for each harmonic in percent of fundamental. That is a fairly typical VFD spectrum for the common six pulse type of rectifier/converter. The question is, does this level cause this drive a problem, or any other load equipment at this point in the facility? If not, then there is enough inductance in the system to "swallow up" the high frequency current.

Next upstream transformer. Are these currents large enough to create a voltage distortion on the secondary bus that effects other devices fed from the bus? If not, then the size of the transformer and its relatively low impedance, compared to what must be low level of harmonic currents, has avoided the danger.

Service point. Are either harmonic currents or voltages in excess of the maximum stated in IEEE-519? If not, then the harmonics have passed the last of the three zones and do not present a problem. The wiring, transformers, bus systems, etc., have dissipated the currents being asked for by the VFD.

While in this example it is not warranted to spend any money for a "fix" at the present, watch out that management does not suddenly "use" the so-called excess capacity to install more load. At that time, there may be a need to examine which is the least expensive alternative for the new capacity. The choices include adding harmonic traps to solve the harmonic mitigation needs, thus freeing up system capacity to do the new work; or investing in new switchgear and apparatus to provide the increased current capacity for the new work. The best answer will be to mitigate the harmonics, free the system ampere capacity at a lower cost than new switchgear, and proceed to handle the new load at the lowest cost possible.

Monitoring equipment is required to locate power problems. Shown is a monitoring device and a current transformer.

Handheld digital oscilloscopes *allow waveform analysis to be carried out.*

TROUBLESHOOTING TECHNIQUES

In previous chapters, the emphasis was on locating and solving system-wide quality power problems that affect solid-state electronic devices. This chapter, instead, will concentrate more heavily on bench and field testing of the devices and their wiring to determine and correct problems that prevent them from interacting well with the system in which they are installed

NOISE IN DATA COMMUNICATIONS CABLE

Electrical noise currents on data communication cables are a real problem. They can cause corruption of the desired signals being sent across the cable by the equipment connected to its ends. In extreme cases, these noise currents may even become great enough to cause electrical damage, such as component burnout, to the circuit elements used at either end of the cable. Learning how to make some simple noise current measurements on such cables without disturbing the signal transmission itself is important if it is desired to troubleshoot such a problem.

Knowing the types of electrical noise and how to measure them are important first steps in enhancing troubleshooting.

Most of the noise currents that get onto a victim data communication cable are of the common-mode (CM) type, as opposed to being the signal [differential-mode (DM)] type. What's the difference between the two?

DM signal current. Look at **Fig. 10.1**. A current can be exchanged between two conductors on either grounded or ungrounded circuits of any type. Such a current is called a differential-mode current. In some cases, it is also referred to by another name; normal-mode (NM), but differential mode is the preferred name.

CM noise current. In **Fig. 10.2**, what is shown is that a current can be caused to flow simultaneously in the same two conductors seen in Fig. 10.1, but with negligible DM potential difference between the two conductors due to this current. This kind of current flow is called common-mode current since it is simultaneously common to both conductors. A CM current typically (but not always) uses the grounding system to complete the circuit loop for its circulation between the victim cable's ends.

DM current on a data communications cable is typically created by the desired action of cable's electronic circuits, which are used at the cable's ends to send and receive the desired signal, whatever it is. CM current is caused by an aggressor source of noise current and by any combination of the following means.

• The ends of the victim data com-

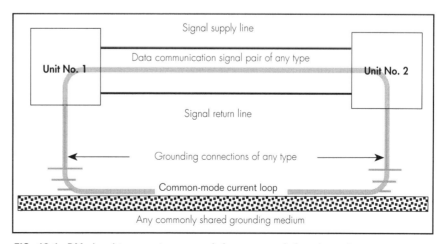

FIG. 10.1. *DM signal transport on grounded or ungrounded equipment.*

munications cable are referenced to "ground" by either a direct connection, terminating electronics circuit impedances, or a stray capacitance of some sort. In addition, the two ends are not at the same potential. So, a conducted CM current flows as shown in **Fig. 10.3**.

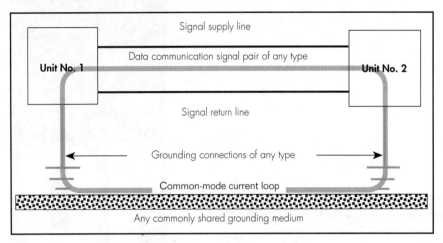

FIG. 10.2. *CM current flow on data communications cable between units.*

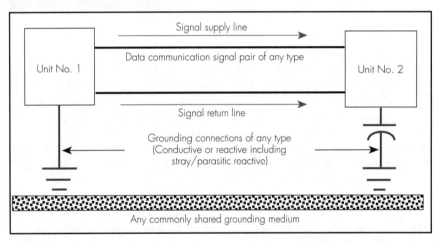

FIG. 10.3. *Potential difference between data communications cable ends causes CM current to flow on cable.*

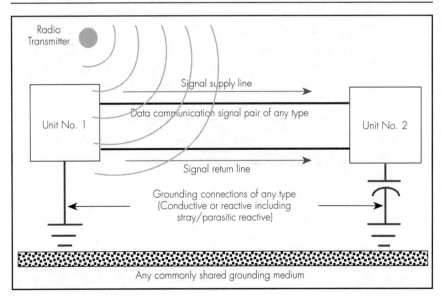

FIG. 10.4. *Data communications cable acts as an antenna to receive radio signals in CM current using ground circuits to complete the current loop.*

• The victim data communications cable looks like a loop antenna and, as shown in **Fig. 10.4**, receives a wide range of radiated radio signals from the air, circulating them in common-mode between the cable's ends via the grounding system.

• The victim data communications cable acts as half of a capacitor's plate and, as shown in **Fig. 10.5**, is electrostatically coupled to any nearby conductor's electric fields. Air is one of the dielectrics between the two "plates" that make up the capacitor.

• The victim data communications cable acts as a single-turn, air-core transformer secondary winding, as shown in **Fig. 10.6**, and is magnetically coupled to any nearby conductor's magnetic fields.

To further keep these descriptive terms straight, note that conducted CM current occurs because the ends of a data communication cable are in some way galvanically (metallically) connected into the grounding system by either a direct connection or via circuit impedance in the cable's terminating electronics.

In contrast, all other CM current is introduced due to coupled means. These latter kinds of unwanted CM currents are due to far fields (radio waves through the air), or near fields (electric or magnetic field) action.

DIAGNOSTIC INSTRUMENTS

Now that the different types of noise are understood, it's time to look at how to make the necessary CM current mea-

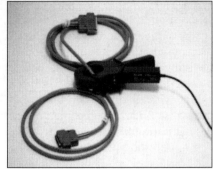

Photo 1. *The CT picks up the CM current because its flow generates a magnetic field around the conductors that is proportional to the current's magnitude.*

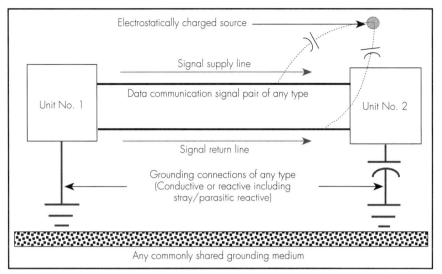

Fig. 10.5. Stray/parasitic capacitance couples electrostatic fields from charged source onto signal pair in data communications cable.

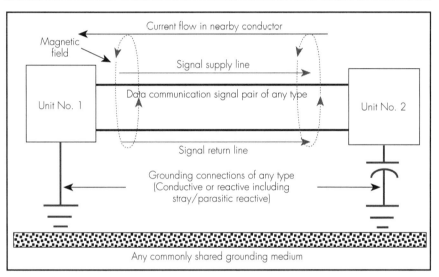

Fig. 10.6. Nearby conductor couples magnetically onto data communications cable, which is acting as a single-turn secondary, air-core transformer.

Photo 2. A simple average responding or true rms type of analog or digital meter can be used to measure CM currents in the range from DC to about 1.5 kHz on a 60-Hz AC system.

surements by using typically available test equipment.

Clamp-on CT. The key to making a CM current measurement is to use a clamp-on current transformer (CT), as shown in **Photo 1**, as opposed to trying to use a series-connected shunt and looking at the voltage drop across it. The shunt is not normally used because it obviously requires cutting into the victim circuit's conductors in order to insert it. Also, its installation can change the circuit's electrical characteristics. A CT can be installed onto a single or multiple conductor cable without affecting the circuit's performance or cutting into the conductors.

A CT picks up the CM current because its flow generates a magnetic field around the conductors that is proportional to the current's magnitude.

DM currents, on the other hand, do not generated very much magnetic field since the signal pair is physically configured so that the magnetic fields cancel each other out. This is called a zero-sequence magnetic field condition, a phenomenon familiar to electricians using clamp-on ammeters when they gather all of a circuit's conductors through the meter's aperture at the same time. Any current that is displayed is called unbalanced current or ground-leakage current, depending upon the situation.

A CT can pick up the presence and magnitude of a CM current, but what is also needed is an indicator.

Analog or digital meter. Typically, if the interest is only in CM currents in the range from DC through about the first 25th or so harmonic of the power system's fundamental frequency (to about 1.5 kHz on a 60 Hz system), a simple average-responding or true-rms type of analog or digital meter is often used, as shown in **Photo 2**. These kinds of meters are quite useful in finding and tracking CM currents on data communication cables, where the problem can be related to the AC system's fundamental of the first few tens of harmonics. Such currents are common in most facilities and appear most often on cables that are suffering from conducted CM noise currents or from near-field magnetic field coupling.

Oscilloscope. The above kinds of restricted-bandwidth detector/indicators are often not useful in finding interfering CM currents that range into the tens of kHz or well into the MHz region. Also, more sensitivity than provided by the typical meter may be

Photo 3. A CT can be interfaced to an oscilloscope so that a broad range of frequency as well as lower level signals may be graphically displayed.

Photo 4. Where measurements are needed well into the MHz range, a wide-band, laboratory grade CT is used. A typical CT as shown here has a bandwidth from around 1 kHz to over 150 MHz.

needed to adequately "see" the CM current. As such, the CT can be interfaced to an oscilloscope, as shown in **Photo 3**, for a simultaneous display of a broad range of frequencies as well as lower level signals. This ability to see the CM current on the time-domain (x= time, y= amplitude) also helps visualize what the CM current looks like (its signature). This can really help in identifying its source and in tracking the same CM current through several points of measurement.

Wide-band, lab grade CT. Where a broad-bandwidth detector is needed, such as where measurements are needed well into the MHz range, a wide-band, lab grade CT is used. A typical CT of this type is shown in **Photo 4**, and is used where the bandwidth is from around 1 kHz to over 150 MHz (other ranges are available).

Hall-Effect based CT will need to be used if CM current below a few Hz needs to be checked. Sensors of this type will simultaneously respond to DC and AC but may not be usable above about 100 kHz. Also, they may be affected by external magnetic fields.

COMMUNICATION CABLE NOISE EXAMPLE

The best way to describe how the communications cable CM noise problem should be analyzed is by describing an actual procedure.

Site conditions. The victim coaxial cable was 52-ohm, RG-58/U cable approximately 50-ft long that ran from a transducer located in the field to a control room via an overhead PVC conduit.

The transducer end was provided with a 2-wire-plus-ground connector, a switch-mode power supply (SMPS), and was fed by a 120V branch circuit. The control-room end terminated with a BNC connector to a proprietary interface card on the back of a PC-type controller.

A preliminary measurement using a portable oscilloscope and a 10× voltage probe on Channel B showed a corrupted signal similar to the waveform shown in **Fig. 10.7** on the input to the cable end's receiver plug-in card. This badly corrupted square wave signal was about 1.8V peak-to-peak, with a typical period of about 2 ms (500 Hz).

Troubleshooting procedure. The coaxial cable was removed from the controller end and a BNC-style TEE fitting attached in its place as shown in **Fig. 10.8**. This allowed a tap to be made across a noninductive 50-ohm resistor, one end of which is soldered to the signal pin of one port of the TEE.

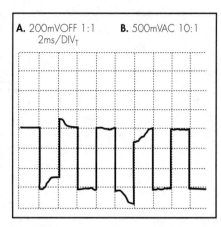

Fig. 10.7. Waveform of corrupted data signal.

The first step was to see if any significant CM current, which could be correlated in some way to the corruption of the signal, existed on the victim coaxial cable itself. This was done by connecting a CT to the portable oscilloscope and then clamping it around the victim coaxial cable.

Fig. 10.9 shows the result of this test.

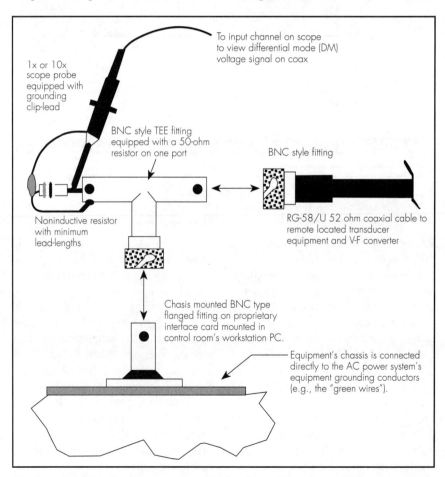

Fig. 10.8. Tap is made across a noninductive, 50-ohm resistor that was previously soldered in place across the signal pin of one port on the TEE to its metal shield/enclosure. This permitted the viewing of differential mode voltage signal on the coaxial cable.

The waveform looked familiar; it was like a typical signature current for the AC input to an SMPS. But an interesting question was: did this CM current have a timing relationship to the EMI on the victim cable's signal?

The next step involved using the voltage probe between the TEE connection and the portable oscilloscope in an attempt to answer this question. This time, however, a two-channel setup was required so that both CM current and data signal voltage could be viewed simultaneously. The result is shown in **Fig. 10.10**. A clear relationship between the EMI and CM current existed.

Next, the TEE connection was disconnected from the rear of the controller but left attached to the B-Channel voltage input on the portable oscilloscope to see the result with the CM-current path on the victim coax cable path completely open at the receiver end. The result was the clean squarewave signal shown in **Fig. 10.11**.

Now the question was, where was the CM current getting onto the victim coax cable to begin with?

Towards the solution. The transducer end provided the answer. Its power supply was disconnected from the branch circuit, and with the TEE fitting reconnected to the controller, another CM current measurement was made. No appreciable CM current was found to be present.

One more test was made by placing the CT around the AC power supply cord for the SMPS to see if CM current could be measured there. It was, and the conclusion can be drawn that the power supply was probably the aggressor source of the EMI.

Bench tests. It now was time to pull the switch-mode power supply and check it on the bench to see what could be found. The SMPS was powered up on the test bench using the adapter shown in **Fig. 10.12**. The arrangement allowed the placement of a CT around any one or combination of the three AC line conductors used in the power cord.

The equipment safety grounding conductor was the first tested. The CT was placed around the green wire and the SMPS was switched on. Except for amplitude, the portable oscilloscope clearly showed a signature current waveform of the same shape and timing as had been observed on the victim coaxial cable. This was clearly not a typical permitted small leakage current. Something was wrong inside the power supply.

A continuity tester was used to check from each of the unplugged SMPS line cord pins to the metal enclosure of the SMPS. This showed good continuity for the green wire, an open on the ungrounded side of the line (the black wire), and a short on the grounded (neutral) side of the line (the white wire).

A closer look inside the SMPS revealed a shorted AC line filter capacitor connected between ground and the neutral pin, as shown in **Fig. 10.13**. Thus, some portion of the SMPS neutral current was dividing between the neutral and equipment grounding paths, which consisted of the green wire and the grounded copper braid of the victim coaxial cable. The controller's input power cord supplied the required connection back to the electrical system neutral of the transformer. The defective SMPS was replaced.

The remaining item that remained to be answered was how the ground-fault current flowing on the coax cable's braid could combine with and corrupt the instrumentation signal. This will not normally happen when coaxial cable is used for signal transport.

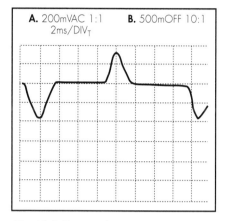

Fig. 10.9. Unwanted common-mode current on the coaxial data cable.

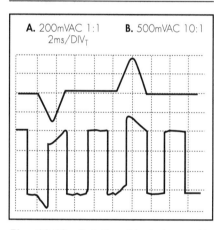

Fig. 10.10. Relationship between the common-mode current on the data cable and the corrupted data signal.

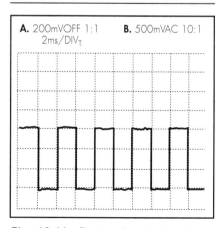

Fig. 10.11. Restored data signal now essentially free from any EMI.

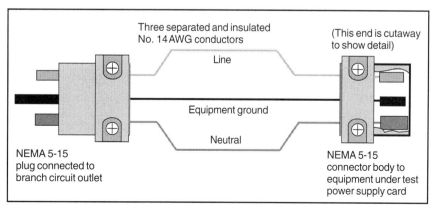

Fig. 10.12. Current transformer adapter cord for NEMA 5-15 circuits.

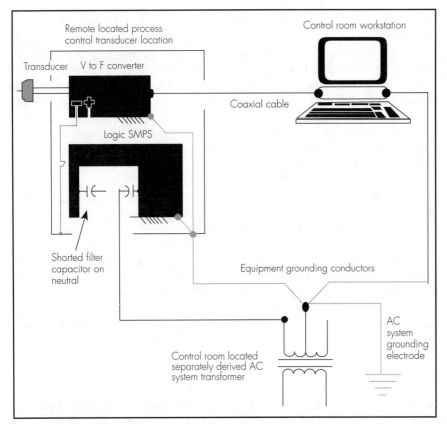

Fig. 10.13. A shorted AC line filter capacitor, which was connected between ground and the neutral pin, resulted in some portion of the SMPS' neutral current to be divided between the neutral and equipment grounding paths.

The answer is that it was very low-frequency signal, which flows on the braid and in the center conductor with very little skin-effect. Without skin-effect, both the desired signal current and undesired currents were free to flow in the entire available cross-sectional area of the braid, similar to DC. Thus, the two currents shared the same conductor volume and were algebraically added together to produce a voltage drop on the cable due to the braid's impedance. This action produced the combined signal voltage in the differential mode at the receiver end of the cable.

If the signal had been of a high frequency, it would have flowed entirely on the inner surface of the braid, nearest the center conductor. This would have left a negligible amount of shared conductor volume for the two current to mix together.

DECIBELS AND ELECTRICAL MEASUREMENTS

The decibel (dB) is used in electrical measurements in innumerable ways. For example, how much of a signal is lost over a transport path such as a coaxial cable or similar metallic path is generally described using decibels. Also, the gain of an amplifier is generally expressed in decibels.

Now, dB is usually directly computed by test instruments such as solid-state oscilloscopes. Still, its best to know what the dB is and how to work with it manually in order to better understand what a test instrument displays when measuring dBs.

Basic units. dB is expressed in relation to the electrical unit it's to be used with. It's important to express dB in this manner to avoid confusion. For example, dBV and dBmV are used for decibels expressed in terms of volts or millivolts; dBA and dBmA for decibels expressed in terms of amperes and milliamperes; and dBW for decibels expressed in terms of watts. Oddly enough, dB expressed in terms of milliwatts is simply abbreviated as dBm.

Prefixing dB with a minus sign (-) means a loss, and either no sign or a positive one (+) means a gain.

General equations. A description of the general equation for determining the dB of gain (+dB) or loss (-dB) for any set of two voltages on a path of equal impedance is:

$$dB = 20 \log (E_1 \div E_2)$$

Solving this equation is not difficult with a calculator that has a "log" key on it. All that's needed is to compute the ratio of the input to output voltage (E_1/E_2), press the "log" key, and then multiply the result by 20. This gives the result in terms of -dB or +dB depending upon whether signal was lost or added to in the circuit.

Current (I) can be substituted for voltage in this equation.

Power calculations are a different matter. It's calculated using a factor of

dB	Power ratio	Voltage or current ratio	dB	Power ratio	Voltage or current ratio
0.0	1.00	1.00	10	10.0	3.2
0.5	1.12	1.08	15	31.6	5.6
1.0	1.26	1.12	20	100.0	10.0
1.5	1.41	1.19	25	316.0	18.0
2.0	1.58	1.26	30	1000.0	32.0
3.0	2.00	1.41	40	10,000.0	100.0
4.0	2.51	1.58	50	$10^5.0$	316.0
5.0	3.16	1.78	60	$10^6.0$	1000.0
6.0	3.98	2.00	70	$10^7.0$	3162.0
7.0	5.01	2.24	80	$10^8.0$	10,000.0
8.0	6.31	2.51	90	$10^9.0$	31,620.0
9.0	7.94	2.82	100	$10^{10}.0$	$10^5.0$

10.14. Listing of common decibel values based on varying power and voltage/current ratios.

Decibels are often thought of as applying to acoustic noise generated by mechanical equipment. In fact, dBs also are used to describe functions in electrical circuits.

10 rather than 20. The basic power equation for computing dB in a circuit of the same impedance is:

$$dB = 10 \log (P_1 \div P_2)$$

Short cuts. Values of dB calculated using the equations based on varying ratios of power and voltage/current are shown in the table of **Fig. 10.14** as a quick reference. A few really important relationships in dB are worth memorizing to quickly estimate something. For example, the following are useful:
• doubling or halving of voltage or current is a ±6dB change;
• a numerical ratio of V or I of: 10 =20dB; 100 =40dB; 1000 =60dB, etc.;
• doubling or halving of power is a ±3 dB change;
• a numerical ratio of W of: 2 =3dB; 4 =6dB; 10 =10dB; 100 =20 dB; etc.

SAMPLE dB CALCULATION

The nice thing about dB notation is that gains and losses in any given circuit are simply added and subtracted arithmetically to find the final value of gain or loss. Then with a little algebra or with the "antilog" key (10×) on the calculator, the voltage, current, or power ratio can be determined from the resulting answer.

Here's how losses of a given signal transport, and amplifier (gain), work together to produce a given output from a specified input.

If the ±dB at any point in a system is known, the voltage or current ratio can be computed using the following equation:

$$(E_1 \div E_2) = 10^{(dB/20)}$$

This is the basic equation for computing the voltage ratio in a circuit of the same impedance when the ±dB is known. To use it, first divide the known dB value by 20. The result is the power to which 10 is to be raised. Using the calculator, push the Xy key. The resulting value is the ratio between the input voltages.

Assume there is a 12dB gain after the signal is "processed and transported" from the source to the load. Using this equation and the procedure described, this 12dB works out to a ratio of 3.98:1. This value × 10 =39.8V output.

The ratio for power is computed in the same manner except that 10 is used in place of 20 in the exponent's divisor, as shown in the following basic equation for computing the power ratio in a circuit of the same impedance when the ±dB is known.

$$(P_1 \div P_2) = 10^{(dB/10)}$$

CCTV SIGNAL ATTENUATION EXAMPLE

Once again, the best way to explain how to apply this knowledge, along with modern test instruments, in troubleshooting any electronic equipment operation problems is by citing an actual example.

A client complained about a very poor picture generated by a security camera on a closed circuit television (CCTV) video monitor on a simple point-to-point link. The picture had degraded from good to unusable over a short period of time and really got bad in the few days following a recent thunderstorm. The general opinion was that lightning had somehow caused the problem. This case history shows how decibel knowledge, a handheld oscilloscope, and intuitive thinking can be used to solve such a troublesome problem.

Initial conditions. The picture on the video monitor looked as though someone had turned the contrast control all the way in one direction so there would be no contrast; the picture was washed-out and barely visible on the screen.

Upon inspection, the CCTV system was found to consist of National Television System Committee (NTSC) standard video generated and displayed in black-and-white. The main monitoring point of the CCTV system was a loading dock. A weatherproof-housed camera located on a 10-ft high post mounted atop the building's roof was used to monitor that location. The camera was connected to a video monitor in a security shack some 200 ft away via a 75 ohm coaxial cable, which was routed down into the shack by means of a pipe-sleeve-type roof penetration. Power to the camera was provided via a low-voltage DC link on another coaxial cable, which was installed right alongside the coaxial cable used to transport the video signal. Both the monitor and camera low-voltage DC supplies were plugged into a wall outlet convenient to the operator at the guard's shack.

Field tests. The first task was to go to the camera at the roof and see what the video signal looked like as it exited the camera. This test was much aided by the fact that our handheld, 50 MHz bandwidth, solid state, digitizing oscilloscope with LCD display had an internal battery pack and did not require any AC power for operation.

First test. The coaxial cable was disconnected at the camera and a BNC style TEE fitting (similar to the one in Fig. 10.8 except fitted with a 75 ohm terminator resistor). The handheld oscilloscope was connected into the remaining open end of the TEE. The result, as shown in **Fig. 10.15**, was a healthy NTSC composite video signal. Conclusion: the camera was clearly putting out a good signal, which was about 1.8V peak-to-peak across the 75 ohm termination. (There is also a DC component with the AC video signal.)

Next, the oscilloscope was switched to its meter-mode and the signal at the

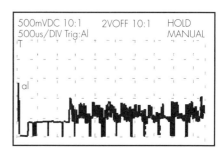

Fig. 10.15. *Proper video signal display taken at camera's 75-ohm output with 75-ohm termination fitting installed in place of existing coaxial cable.*

camera into the 75 ohm load was taken as a zero dB reference and stored into memory. This is shown in **Fig. 10.16**, where +000.1 dBV DC is taken as being close enough to zero to do the job. Now this good camera signal was available for use again and again as a comparison with signals measured at different locations. Thus, it would be possible to see how much loss of signal occurred along the path, all of which was supposed to be a consistent 75 ohm.

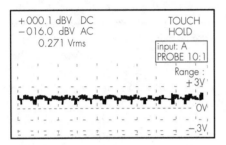

Fig. 10.16. Output video signal measurement, as taken at camera with 75-ohm termination fitting in place instead of existing coaxial cable, is shown in handheld oscilloscope's meter mode. This dBV measurement was used to establish the zero dB level of the video signal for referencing and subsequent signal-loss measurements.

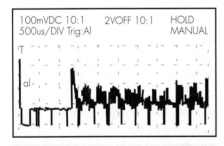

Fig. 10.17. A scale change from 500V/cm to 100V/cm on the handheld oscilloscope enabled the viewing of this degraded video signal at the monitor end of the 75-ohm coaxial cable.

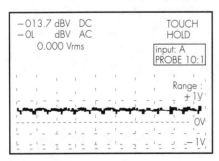

Fig. 10.18. This dBV measurement, made with the handheld oscilloscope in its meter mode, was taken at the monitor end of the 75-ohm coaxial cable. The signal loss is -13.7 dBV in 200 ft.

The TEE was removed and the 75 ohm coaxial cable was reconnected.

Second test. This test was made at the video monitor end of the cable and right at the point where the cable was connected to the monitor. Again, the TEE was used, but this time no 75 ohm termination resistor was used with it since the TEE was attached to both the monitor and the cable. Thus, there was a fairly good 75 ohm load on everything. The result of this test was that almost no video signal could be seen on the oscilloscope's screen.

Next, the oscilloscope's vertical scale was changed from 500 mV/cm to 100 mV/cm and another measurement was taken, which is shown in **Fig. 10.17**. The video signal is simply attenuated but does not appear to be distorted in any way that is easy to see. Conclusion: the signal loss was occurring along the 75 ohm cable path.

A quick test with the TEE and the 75 ohm termination resistor in place of the video monitor quickly ruled out the possibility that the video monitor was "loading down" the signal due to an internal failure on its input circuit. The signal, however, was essentially unchanged from that shown in Fig. 10.17. This confirmed the signal's loss was occurring along the 75 ohm cable.

Third test. The oscilloscope was placed into its meter-mode, while maintaining the connection to the TEE at the junction of the cable and video monitor. This allowed taking a relative dB measurement reading, as shown in **Fig. 10.18**, using the original zero-level as the reference. Here, a -13.7 dBV DC loss can be seen to exist. This loss represents a voltage loss ratio of 4.84:1, or a signal loss of nearly 5V for every volt put into the cable!

Checking manufacturer's data. How much signal attenuation should be expected on a 200-ft long, 75 ohm coaxial cable? A quick look at the coaxial cable manufacturer's Master Catalog gave the approximate answer: around 2 dB of loss at 10 MHz for 200 ft of RG-59/U type cable as used in CATV applications. The whole attenuation chart is shown in the accompanying table of **Fig. 10.19**.

What was actually seen in this path was more than 11 dB loss over and above that stated in the manufacturer's literature. Also, the baseband video shouldn't have a lot of really high-frequency in it; thus, the cable probably shouldn't attenuate as much as 2 dB (per manufacturer's literature) for 200 ft in any case.

Oh yes, since the manufacturer's information was provided only in dB form, what could have been done without an understanding of dB and working in terms of dB on the oscilloscope? You guessed it. We would have had no idea what was normal and what was not on a coaxial cable run of the type being investigated. All would have been guesswork, which is not a very good way to troubleshoot.

Further analysis. What was happening on the cable? An examination of the BNC at the camera end failed to show any problems. The connections at the camera also were thoroughly checked to make sure they were well protected from the environment by the enclosure.

On the other hand, the BNC connector at the video monitor end was inspected and it seemed to be a little wet after it was handled and the cable flexed. This moisture was a clue. Past experience with coaxial cables with water inside showed this condition caused severe signal attenuation.

A closer examination of the 75 ohm cable was made from the camera back along its route to the video monitor. This lead to close inspection of the rooftop penetration through which the cable was passed. It was not equipped with a weatherhead; the cable was simply stuck down into it from its open top. Sealing was done with a putty or caulk-

MHz	dB/100 ft	dB/100 m
1	.3	1.0
10	.9	3.0
50	2.1	6.9
100	3.0	9.8
200	4.5	14.8
400	6.6	21.7
700	8.9	29.2
900	10.1	33.1
1000	10.9	35.8

Fig. 10.19. Nominal attenuation of a typical CATV RG-59/U type 75-ohm coaxial cable.

ing material and it looked dried-out. Thus, water from the storm could follow the cable sheath down into the building around the bad seal.

But how did this condition let the water into the cable? The sealing material was removed from the penetration pipe and the coaxial cable pulled out. About 10 feet down, a connection was found made up of two BNC fittings and a male-male adapter. The end of the cable going into the bottom BNC fitting from the building was partially pulled out of the connector and the braid/sheath was fully exposed to any water flowing down the cable from above. In fact, the arrangement was a good funnel for the water to flow into the cable between the outer sheath/shield and the inner dielectric material. Corrosion was also rampant in the damaged connector set since it had not been sealed from moisture in any way. Obviously, this was not good for reliable signal transport.

Where did this splice come from? After a little discussion with the personnel, it was learned that the camera came from the factory with about 10 ft of cable. Rather than throw this cable away, it was kept in place and used by connecting it to the end of the cable routed from the video monitor. There was no explanation as to why such a poor rooftop penetration was made; nobody would own up to it.

Solution. The whole existing 75 ohm coaxial cable run was replaced with a continuous length one. After installation of the new cable, the signal at the video monitor end was again checked with the test TEE and the monitor in

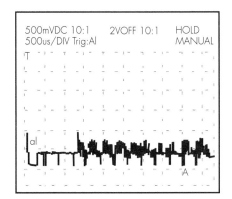

Fig. 10.20. *Display of video signal with "new" coaxial cable installed and test TEE at the monitor in place. This signal produced a very acceptable picture.*

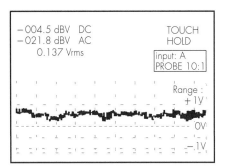

Fig. 10.21. *This dBV measurement, made with the handheld oscilloscope in its meter mode, was taken after installation of "new" coaxial cable. It shows an acceptable -4.5 dBV signal loss from the camera to the monitor.*

place. This signal is shown in **Fig. 10.20**. There is still some attenuation, but nowhere near as much as before.

Again, using the handheld oscilloscope in its meter-mode, a relative dB measurement reading was made using the original zero-level as the reference. The new cable's signal loss, as shown in **Fig. 10.21**, was about -4.5 dB. Compared to the previous dB measurement readings, this amount of signal loss is acceptable in this application, as was evidenced by the good picture on the video monitor.

ISOLATION TRANSFORMER ATTENUATION EXAMPLE

How can electrostatic-shielded power isolation transformer performance in attenuating common-mode electrical noise be verified? Here's a down-to-earth test method that resembles real-world applications. All that is needed is a special (but simple) test stand arrangement, which involves a signal generator, a standard PC power supply, a shield in/out switch, and an oscilloscope. The results will adequately represent the transformer's performance.

The need is to find out how much common-mode current the transformer's shield will keep from reaching the victim load. There's no interest in finding out how little common-mode voltage can be developed across an OEM-selected test-stand capacitor. This device is unrealistically substituted in factory tests for the solid grounding conductor required to be installed per NEC Sec. 250-26 on those separately derived AC systems with grounding required by Sec. 250-5.

With this in mind, look at **Fig. 10.22**. With this test setup, it's possible to find the amount of common-mode current the transformer really delivers to a typical victim load with the electrostatic shield both open and grounded across a ground plane.

Test setup. A 386/486 PC switchmode power supply is used as the victim load in this example. As shown, a CT is place around the grounding/bonding jumper, between the PC power supply's common terminal and the overall test stand ground reference. This ground reference is established by a sheetmetal ground plane. This setup approximates real-world practice for most information technology equipment.

A very high-quality, 500VA electrostatic shielded isolation transformer was used in this example. A handheld os-

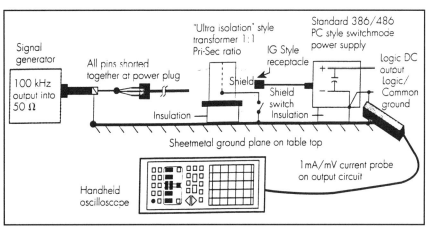

Fig. 10.22. *Test stand arrangement to measure common-mode current attenuation includes a signal generator, a standard 186/486 PC power supply, a shield in/out switch, and a handheld oscilloscope.*

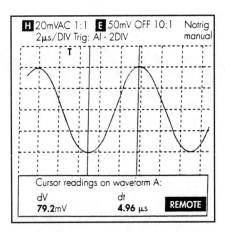

Fig. 10.23. Logic power supply grounding/bonding strap's current at 1mV/mA and 20mV/cm, at 100 kHz, and with the electrostatic shield ungrounded.

cilloscope was used having a broadband current transformer that outputs 1 mV per mA into the oscilloscope.

Testing is performed at two frequencies: at about 100 kHz, where lightning contains a lot of energy, and at 20 kHz, the stated MIL-T27(D) production line test frequency.

Testing at 100 kHz.

With the shield open, the amount of current at 100 kHz, as shown in **Fig. 10.23**, is measured at 79.2 mA. This is the amount of common-mode noise current being passed through the isolation transformer's interwinding capacitance with the electrostatic shield ungrounded, or not in the circuit. This test result shows the noise current neatly goes around the PC power supply, as opposed to going through it, to get to the logic DC buses. This initial condition is established as the zero dB reference point from which the effective performance of the shield can be compared once it is grounded.

With the shield grounded, and no other changes to the test stand other than to ensure the same amount of current is being delivered by the signal generator to the isolation transformer's input power cord, the shield is low-inductance grounded to the ground plane. The current, again, is measured in the power supply's grounding strap. As shown in **Fig. 10.24**, there is a significant reduction in common-mode noise current, down to 8.8mA.

Calculating attenuation. But, how much reduction is this really? To find out, it's necessary to go through the typical calculation, with the result expressed in terms of -dB (attenuation loss).

A comparison can be made between the first test result, at 79.2mA, to that from the second test, at 8.8mA, and the loss or attenuation in dB can be determined by using the standard equation: dB = 20 log(E_1/E_2). This results in an attenuation of only -19dB of common-mode current. Another way of saying the same thing is: a common-mode current attenuation of 9:1 exists.

Test at 20 kHz produces a common-mode current attenuation of -22.3dB with the test currents of 130mA and 10mA with the shield open and grounded, respectively. This is an attenuation ratio or 13:1.

There's a little over 3 dB of difference between the two test results; however, this doesn't take into account the fact that the attenuation curve typically may not be smooth between the two test frequencies used.

Testing existing installations. The above procedure can be used on existing isolation transformers installed in the field at real installations to determine the amount of common-mode current attenuation of the arrangement. To do this, the common-mode current is measured on the output circuit from the isolation transformer's secondary by placing all the conductors and their raceway into a toroidal form or wideband current transformer instead

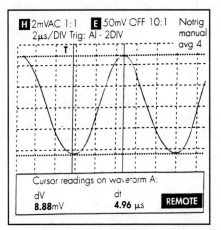

Fig. 10.24. Logic power supply grounding/bonding strap's current at 1mV/mA and 2mV/cm at 100 Hz, and with the electrostatic shield grounded.

of using a small-sized intrumentation probe. Thus, the total common-mode current conducted to/from the transformer's secondary to its served loads can be measured with the shield open and grounded.

EXAMPLE, LOCATING AN EMI SOURCE

A rack of sensitive electronic load equipment was installed in a secure area at an aerospace facility. Also in this area was some equipment generating high-level electromagnetic interference (EMI). The EMI emissions were negatively affecting the electronic load equipment. A quick inspection showed that the victim equipment's low-level logic wiring had the EMI signal all over it. EMI was also found at fairly high levels on the AC power circuits in the secure area as well as on the branch circuit feeding the victim rack.

Because the site was classified, no modification of the secure area or the classified equipment was permitted. Only the victim equipment could be modified, due to its temporary testing usage.

The victim equipment consisted of typical rack-mounted electronic devices. Low-level signal cables ran between each device and were routed vertically in the rack's left-rear corner. High-level signal cables were similarly interconnected but routed in the rack's right-rear corner, near a 120V plug-in strip. AC power to the rack was via an unshielded Type SO flexible cord equipped with a connector at each end. The 10-ft long cord had approximately 7 ft of excess length coiled up and stored near the rack under a raised cellular floor.

As part of the classified wiring design, a heavy gauge, copper wire grounding conductor system was installed under the raised cellular floor. All equipment within the secure area was bonded to this system with very heavy gauge, tinned copper braid straps. One of these straps was connected to a large metal stud on the outside face of the victim equipment's AC power input panel.

Test 1. The first test involved connecting a dual channel oscilloscope, in the differential mode, to the victim rack's line-to-neutral branch receptacle circuit. This would show if the EMI was being conducted. The scope showed a large amount of EMI with characteristics almost identical to that being produced by the classified equipment. It also showed that the EMI impulse's first transition time was in the megahertz (MHz) range.

Conclusion: Since the AC power source was a 225kVA transformer with 4% impedance, the interference probably was not the result of load-source power supply interaction across a commonly shared impedance. And since the EMI impulse's first transition time was in the MHz range, the EMI looked more like an unwanted signal being emitted from the classified equipment, on its own branch circuit, and conducted onto the secure area's AC power circuits.

Test 2. Radiated EMI still could not be ruled out. To verify this, a 50-ohm terminated loop (sniffer) antenna, made of coaxial cable was connected to the oscilloscope to see if anything would be picked up. The sniffer antenna was used in the general area of the victim rack and easily picked up the subject EMI signal. Turning the sniffer antenna to find signal nulls (areas of signal attenuation) showed that the EMI signal was definitely coming from the classified equipment. However, when the sniffer was directed towards the victim equipment, a radiated EMI signal was also picked up. Both signals looked the same on the oscilloscope. Conclusion: the same EMI signal was being simultaneously radiated from two places. Placing the sniffer antenna into the victim rack caused a very large EMI signal to be seen on the scope.

Test 3. To see what portion of the subject EMI signal was being radiated, the victim rack was disconnected from the outside world by removing its AC power cord connection and all signal cable connections. The oscilloscope was then well bonded to the victim rack with a short jumper and the sniffer antenna was again placed into the now-isolated victim rack. (Note: The scope's AC power cord and the sniffer antenna's coaxial cable were both equipped with ferrite core common-mode choke coils.) EMI signal levels would then be checked with the victim rack's door open and closed.

With the door open, a high level of EMI showed on the scope. With the door closed, the EMI almost disappeared. This was surprising since the rack was not a metal shielded type and not fully EMI gasketed. From this, we concluded that simple shielding would not solve the problem. Now it looked like this might be exclusively a conducted EMI problem. But from what path?

Conflicting opinions. At this point, site personnel did not agree with the conclusion of a conducted EMI because the victim rack was equipped with an expensive form of an LC low-pass filter in its AC power interface chassis. All AC power was supposed to pass through this filter before going further into the victim rack. Also, the filter's cut-off frequency was way below that of the EMI, which would result in an EMI attenuation of at least 150 or 160 decibels (316:1 and 1000:1 respectively). Such an attenuation, argued the site personnel, should keep the victim rack's electronics free from EMI problems.

Test 4. Because the conflicting opinion, an additional test was done. This time, the victim rack's AC power cord was reconnected and the sniffer antenna test was run again, with the rack's door opened and closed. Now, even with the door closed, a large EMI signal was being picked up from the rack. This was interesting since the rack's 2-pole input AC circuit breaker was open. Closing the breaker caused the EMI in the rack to further increase, to levels seen during the first sniffer antenna test with both AC power and signal cables attached.

Physical evidence. Because of the results of our last troubleshooting test, we removed the AC power input chassis from the victim rack and examined it. The following items were seen.
• The whole panel is unshielded; no 6-sided metal box fully enclosed it.
• The LC filter's input and output wiring was laced together with plastic cable ties.
• The wires on the panel created fairly large, enclosed open-loop areas, which couple EMI.
• Long cable leads, which make good antennas, were used from the AC power input connector to the LC filter input terminals.
• The 2-pole input breaker was located at the output side of the LC filter.

Conclusions. How do the above items explain the perplexing condition? First, this wiring arrangement allows the EMI to be conducted into the victim rack and then radiated and coupled around the LC filter. Closing the AC power input breaker simply connects the conducted EMI into more wiring inside the rack. From here, increased radiation and coupling occur because of the presence of long cable leads and large open-loops.

In other words, the LC filter is rendered useless and any EMI entering the rack via the AC input power path has full access to the rack's interior. This explains why EMI was present regardless of the position of the breaker. Even with the breaker in the open position, some internal wiring is always connected to the conducted EMI source at the AC power input connector. Simply put, the wiring methods and layout are very poor for the required application.

But, how was the EMI getting into the signal cables and circuits? Usually, low-level circuits are the more likely path for EMI to enter other circuits. As such, we examined the low-level signal cables at the back of the victim rack, looking to see how unwanted coupling between these cables and the EMI source occurred. We noticed that the cables were vertically routed with the victim rack's 120V fan wiring, which, in turn, was fed directly from the AC power interface chassis. The 120V fan wiring was unshielded, untwisted, and not fitted close to the rack's metal surface. Thus, the wiring would be a good EMI source along its entire length. The shielding on the low-level signal wiring couldn't be expected to attenuate enough EMI from such a strong and nearby source.

As the EMI entered the low-level signal path, it was amplified by some of the circuits being used, combined with the desired signals, and distributed to other areas within the victim rack via the high-level cable system. Thus, the EMI got into most everything.

Confirmation. To confirm this, we disconnected the AC power cord and reconnected the low-level signal cables. Using the loop antenna within the rack as we did previously, we noted only a slight increase in EMI level over what was seen when the signal cables were disconnected. We could then conclude that the principal EMI problem was conducted emissions via the AC power circuit.

We noticed that the 120V plug-in strip on the back of the victim rack was nonmetallic. Taking it apart, we noted that the interior phase, neutral, and ground wires were separated by about one inch. Therefore, these conductors, in addition to the fan wires, would represent another source of radiated and coupled EMI within the rack.

The AC input power chassis rack panel was also examined to see how it was grounded. We found that it was poorly grounded because it was painted and no effort was made to ensure good, bare metal-to-metal contact between it and the victim rack structure. The only contact was via the panel's mounting screws.

Connection of the rack to the classified ground wire was via the classified jumper and then through the panels's mounting screws. Therefore, this ground connection also was not well made.

Recommendations. The following were our recommendations.

• Clean the paint off the AC power interface-rack panel where it mates to the rack's mounting rails. Install EMI strip gasketing to enhance contact along the rack panel door's perimeter. Inspect the rest of the victim rack for similar openings and areas of poor contact and treat them accordingly.

• Install and mate an all-metal box shield to the metal rack panel to fully enclose the LC filter and all of its input wiring. This shield should be unpainted and screw-fastened at frequent intervals.

• Relocate the LC filter's output terminals outside of the box shield. To ensure safety, install plastic covers over all exposed electrical terminals at the rear of the panel.

• Relocate the AC power input circuit breaker to the input line side of the LC filter so that when the breaker is open, no radiated or conducted EMI path is available.

• Change the 120V fan wiring to braided, shielded, twisted-pair cable and route it away from all signal cable circuits. Keep the cable close to a ground plane.

• Install a metallic plug-in strip and keep all signal wiring away from it. Ensure the strip's metallic housing is well and frequently bonded to the rack's metal.

• Keep the victim rack's door closed at all times.

This type of EMI problem is unusual found in equipment that is not mass-produced and therefore not subjected to Nationally Recognized Testing Laboratory (NRTL) product safety testing and related FCC interference testing and certification as a complete system. The field mating of various unrelated equipment coupled with the use of rack enclosures and wiring, all of which have not been electromagnetic capability (EMC) tested as part of an overall system, further complicates the problem. What results is a greater EMI risk and generally poor equipment performance.

DESIGN AND INSTALLATION CONSIDERATIONS

Providing quality power for sensitive electronic equipment involves more than the selection of hardware. Even the best and most costly items purchased with the intent of protecting the sensitive loads can be defeated by poor design and installation techniques. It is, thus, necessary to view UPS systems, harmonic filters, etc., as only part of the solution.

This chapter will offer an insight into the strategies needed to implement an effective clean-power program. Where necessary to clarify the issue, a case history will be given.

INTERACTION AND VARYING IMPEDANCES

Most locations simply do not have just 60 Hz loads anymore! **Fig. 11.1** is a table that shows the spectrum of frequencies in current flowing in a typical incoming power feeder, where the largest portion of equipment served consists of solid-state electronic devices. The spectrum can be characterized as a shopping list from the combined loads, each asking for a little 180 Hz, some 300 Hz, a small amount of 420 Hz, a bit of 540 Hz, all to go along with the supply of 60 Hz. As such, the total resultant current is distorted. In other words, it does not have 60 Hz sine wave characteristics exclusively. How does this phenomenon affect transfer or paralleling operations of electrical distribution systems?

When various harmonic currents flow through an upstream impedance, voltages will be generated at each harmonic order. Depending on the size of this impedance and its capacity to cope with the harmonic demand, either a small or large voltage distortion will occur, and it will be present on the entire electrical bus common to the particular impedance device (transformer, generator, etc.). Thus, interactions from harmonic voltage distortion will occur at all points served by that electrical bus.

When the upstream electrical distribution system is large yet has a low impedance, the harmonic voltage dis-

Fundamental	60 Hz	600A
3rd	180 Hz	180A
5th	300 Hz	240A
7th	420 Hz	120A
9th	540 Hz	60A
11th	660 Hz	30A
13th	780 Hz	20A

Fig. 11.1. Table showing a frequency spectrum for current flowing in a typical incoming power feeder, here the largest portion of equipment served consists of solid-state electronic devices.

tortion will be very low because the system is able to swallow up the relatively small portions of harmonic currents contained in the total current demand. Conversely, if the upstream impedance is high, the harmonic voltage distortion will also be high because the distribution system is less able to accept or cope with the harmonic currents.

Taking this one step further, consider presenting such a harmonic current shopping list to one source having a specific impedance, and then transferring to another having a sizably different impedance. This is exactly what happens with automatic transfers of large nonlinear loads to standby or emergency power sources, or when operating cogeneration equipment in parallel with the electric utility. The sizable differences in impedances allow the harmonic currents present to create voltage distortions, which may be different in magnitude by 10 to 12 times! The result: inhibiting of transfer or paralleling operations.

Case history 1. A large hospital complained about harmonic distortion on its power distribution system. After loss of normal power, startup of the emergency generators, and return of normal power, the problem was so severe that retransfer could not take place to normal utility power.

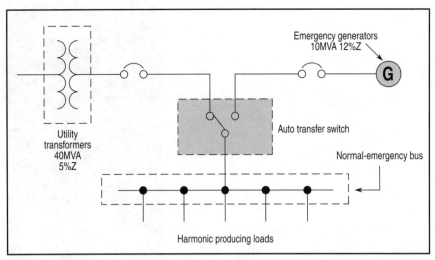

Fig. 11.2. Simplified single-line diagram of hospital's distribution shows the emergency generator's impedance vs that of the utility is almost 2½ to 1, and a disparity in capacity of 40MVA vs 10MVA. These combined differences make the resulting impedance difference almost 10 times more on a per unit basis. Thus, harmonic nonlinear loads create significant voltage drops at the respective harmonic orders and correspondingly high voltage distortion on the generator bus.

The utility provided 40MVA capacity for the hospital, in the form of transformers having approximately 5% impedance. The hospital also had 10MVA of standby generation capacity, with the generators' output impedance close to 12% (**Fig. 11.2**). Upon measurement of harmonic distortion in the current drawn by nonlinear loads connected to transfer switching, a very wide current spectrum was noted.

When these currents were supplied from the utility (normal) feed, there was no excessive voltage distortion. In other words, the extent of voltage distortion present was small relative to the size of the power source and thus was kept within acceptable distortion limits (5% total harmonic distortion on voltage at the point of common coupling with the utility).

On transfer to emergency generator power, the distorted currents were supplied by the generator output, which had a considerably higher impedance than the utility. Specifically, the impedance of the generator versus that of the utility was 12% versus 5%. In addition, there was a disparity of available capacity: 40MVA for the utility versus 10MVA for the generators.

These combined differences made the resulting impedance difference, on a per unit basis, almost 10 times more (12/5 × 4). As a result, the harmonic currents created significant voltage drops at the respective harmonic orders, with a corresponding high voltage distortion on the generator bus. Being sine wave producers, the hospital's generators had a difficult time supplying the distorted current waves resulting from the shopping list of harmonic currents. Nonetheless, the generators accepted the load. The problem occurred on retransfer.

When the utility power returned, a synchronized return to utility power from the emergency power was initiated. When the utility protective relay system saw the high voltage distortion on the bus, it determined the voltage waveform was out of specification for utility service and did not allow the retransfer. The only way the hospital could retransfer to normal power was to turn all loads off while they were still on emergency power, clear the bus, make the retransfer, and then start the loads on utility power. This scenario defeated the purpose of an automatic transfer scheme.

The solution was to limit the harmonic currents at the source locations. This would return the general shape of the total current close to that of a sine wave. Concerns about impedance differences then would be eliminated.

Case history 2. A facility had a new cogeneration installation with 12MVA of capacity and 12% impedance, as shown in **Fig. 11.3**. The electric utility provided 30MVA of transformer capacity with 5% impedance to the site. With all the utility-approved synchronizing equipment in place, testing was started.

During no-load testing, the paralleling operation performed as required. During load-bank acceptance testing, the system again performed as expected.

With the facility's actual loads on line, the testing continued. Now, the nonlinear loads began to show their resulting negative effect: severe voltage distortion on the main electrical bus. This problem appeared, however, only when the equipment load was con-

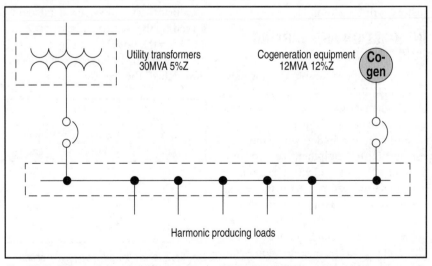

Fig. 11.3. Distorted currents interact more with higher impedance of cogeneration units than with utility system's lower impedance. The resulting distortion is not permitted to be transmitted upstream onto the utility system. Approval of parallel operation is denied until the facility system's harmonic content is reduced to acceptable levels.

nected to the operating cogeneration system. When the equipment load was on utility power only, the problem disappeared.

What happened was the distorted currents were interacting more with the higher impedance of the cogeneration units than with the utility system's lower impedance. The utility wanted to approve acceptance of the cogeneration power but could not permit the resulting distortion to be transmitted upstream onto the utility system. As a result, it denied approval of parallel operation until the harmonic content was reduced to acceptable levels.

Why didn't this problem surface under load bank testing? The answer: load bank testing normally uses linear, resistive loads.

EFFECT OF IMPEDANCE ON THD

The amount of voltage distortion will increase when the source impedance is increased, even when the same currents are flowing. And this increased voltage distortion will be in direct proportion to any multiple of source impedance increase. In other words, the amount of voltage distortion that will occur varies with magnitude of the source impedance when distorted currents are flowing through this source.

Case history. An engine test facility had two large, 6-pulse, adjustable speed drive systems. One drove a 660 hp dynamometer for conducting braking and running tests on a truck with a given engine size. The other powered a 440 hp fan assembly for blowing air over a truck's front end to verify aerodynamic design. Aside from these, there were virtually no other loads, except for a small amount of lighting.

As shown in **Fig. 11.4(A)**, the original service consisted of a 1500kVA, 12.47kV-480V, utility-owned transformer with 6% impedance. The adjustable speed drives produced 5th, 7th, 11th, and 13th order harmonic currents.

An initial power system harmonic analysis showed that the total harmonic distortion (THD) on the voltage was well within guidelines given in IEEE 519. The harmonic voltage distortion at the secondary of the utility transformer was less than 5% of the total of all harmonics, and less than 3% on any single harmonic. And this was with 1100 hp worth of adjustable speed drives producing relatively high percentages of harmonic currents.

Through several billing periods, the utility found the total demand load for the metered facility was only 315kW. Reviewing the facility loads, the utility concluded that, being a regenerative device, the dynamometer was not drawing power most of the time and was instead feeding power back into the system. As a result, the utility determined that the billing was accurate.

Believing it had oversized the service transformer, the utility replaced the original with a 500kVA unit, as shown in **Fig. 11.4(B)**. Its reasoning was simple: with a maximum of 315kW demand load, 500kVA was quite suitable.

Within a short period of time, significant problems resulting from the interaction between the facility power system and the utility's transformer appeared. We examined the situation and found the THD at the point of common coupling (PCC) was now in excess of 12% on the 5th harmonic alone, and over 9% for the entire power system, well above the guidelines.

What had happened was: as the transformer size was reduced, the "apparent" impedance was increased in the same proportion. In other words, the transformer capacity was reduced by a factor of three (1500kVA down to 500kVA), resulting in a new "apparent" impedance of 18%. Now, those same harmonic currents were flowing into what appeared to be a very high impedance source (the 500kVA transformer), causing the voltage distortion to increase significantly.

Our advice was simple: put the original transformer back, replacing the "high impedance" with a much lower one, thus reducing the amount of THD.

Another possible solution was to install a harmonic trap (shown dotted on the 500kVA transformer's secondary) to remove the distorted currents from the line. Without the distorted currents, there would be minimal voltage distortion.

CAPACITANCE AND RESULTING RESONANCE

Pure capacitance cannot be added without running the risk of creating resonance and voltage distortion on existing electrical distribution systems. In addition, the increasing capacitance,

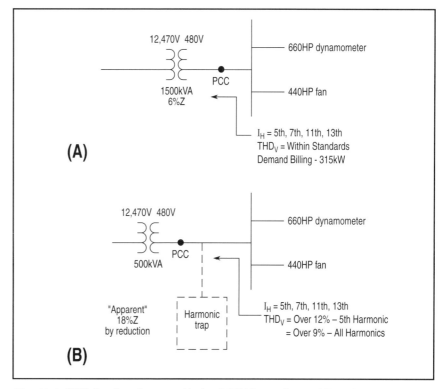

Fig. 11.4. NEW One-line diagram with the 1500kVA transformer (A); and with the 500kVA transformer.

Installation of capacitors for power-factor correction can cause serious problems when harmonics are present. Here, reactors and capacitors are arranged into a harmonic filter bank to avoid resonance in the system.

together with resonance and high frequency currents, will create specific motor problems related to the 5th and 7th harmonic.

In order to insure that such events do not happen, power factor improvement in an existing facility should be approached in the following steps.

• Estimate the capacitance necessary to improve the power factor to the desired value;

• compare the amount of pure capacitance against the rule-of-thumb that says the maximum amount of pure capacitance should be no more than 20% of a transformer's capacity; then

• divide the amount of pure and "protected" capacitance as follows: 3/4 protected, 1/4 pure.

Case history. A limestone quarry had considerable heavy power equipment for crushing and screening. The utility served the facility with 22.9kV power through several unit substations on the site. The owner decided to add a new 450 hp, 6-pulse, adjustable speed drive to a crushing and screening process. (The facility already had considerable heavy power equipment and capacitance on line.) As seen in **Fig. 11.5**, the drive was fed from a 1500kVA substation, which also powered other large motors.

Intent on maintaining power factor as close to unity as possible, the owner also installed 450 kvar of pure capacitance along with the drive. Because the utility's billing was based on kVA demand, it was felt that good power factor levels would result in the least expensive energy costs for the facility. When done, however, without considering transformer sizing and the consequences of power distortion, all sorts of strange interactions can take place. In this case, the result was what might called "self resonance" of an almost pure L-C circuit, and high voltage distortion during operation.

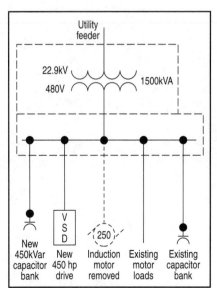

Fig. 11.5. One-line diagram of the substation to which the 450 hp drive was connected.

When we arrived at the site, we took readings at the 1500kVA substation to characterize the currents that were flowing. A harmonic analysis on the 450 kvar capacitor showed that its current consisted of 55% 5th harmonic, 30% 7th harmonic, 25% 11th harmonic, and 10% 13th harmonic (in terms of percent of fundamental current).

Next, we went to the new drive and measured current distortion in somewhat the same proportion: 30% 5th harmonic, 17% 7th harmonic, 10% 11th harmonic, and 5% 13th harmonic.

Based on the above data, we advised the owner that the 450 kvar of pure capacitance should be in the form of a harmonic trap. The pure capacitance was attracting the high frequency currents produced by the input to the drive, and potentially other loads on the substation bus as well. As a result, the capacitance was showing an overload of its total current due to the presence of these high-frequency currents.

Just as a matter of curiosity, we went to the 22.9kV primary service entrance to see if we could identify any harmonic currents coming from the utility. We used a 480V CT on the harmonic analyzer on the secondary side of the primary CTs. Even though the accuracy of the CT applied on a 800:5 ratio primary CT would not be very good, if utility-generated harmonics were present, they would appear in proportion to how they were being supplied on the primary system. Sure enough, we found that the primary service current had the following makeup: 35% 5th harmonic, 20% 7th harmonic, 10% 11th harmonic, and 5% 13th harmonic.

This told us the process equipment on the secondary bus was demanding a distorted wave shape that could be sufficiently generated only by harmonics from a "large" supply, namely the utility. The owner was actually purchasing portions of the resultant distorted wave from the utility so the equipment could operate. Had the pure capacitance been in the form of a harmonic trap, this trap would have "supplied" the high frequency components needed by the load.

We progressed upstream on the utility's system to see if there was any other harmonic source and to determine if the harmonics were being dissipated at some distance away from the quarry. We found at points where we could measure the response of the system, which were several miles away from the quarry, the utility-generated harmonics had completely disappeared. This was an indication of the massiveness of the utility system and its ability to swallow up these currents while still supplying the quarry's load demands.

We came back to the site and conducted tests while changing the makeup of the quarry's load distribution system. First, we removed the newly added 450 kvar capacitor bank from the secondary of the 1500kVA substation. Then we returned to the secondary CT on the 22.9kV side to measure the harmonics again. While the harmonics at the drive remained as noted previously, the high voltage current harmonic composition had changed to the following: 15% 5th harmonic, 12.5% 7th harmonic, and 5% 13th harmonic. This indicated the newly added capacitance was not helping but was, in fact, increasing the amount of current distortion on the system.

Our next step was to remove all the loads from the system, but in a very precise manner. First, we separated the new 450 hp drive from the rest of the loads and, this time, took measurements at the high voltage side as well as at the newly installed capacitor. The capacitor harmonic composition was 25% 5th harmonic, 15% 7th harmonic, 10% 11th harmonic, and 5% 13th harmonic. At the high voltage CT, the composition was 20% 5th harmonic, 10% 7th harmonic, 5% 11th harmonic, and 5% 13th harmonic. In other words, we had verified a resonance condition occurring between the inductance of the transformer and the pure capacitance of the 450 kvar capacitor bank.

We recommended that the owner replace the 450 kvar of pure capacitance with at least a 300 kvar harmonic trap and allow for 150 kvar of newly installed pure capacitance to remain bus connected.

The owner then told us of another problem that had been experienced. An existing 150 hp induction motor, which was only a year and a half old, had to be removed and sent to a repair shop, where it was found to be beyond repair. The question was, why would the damage almost appear to have been caused by a bomb blast?

The answer was: it could be expected because of the existence of distorted voltage. The 5th harmonic voltage in a 3-phase system is a negative sequence voltage, and will produce an opposing torque. As a result, the motor will not have enough torque to perform its work and will call for the electrical system to send more current to produce more power. In the process, more and more heat is generated until the motor fails. We told the owner the same situation would occur with every replacement motor installed until the 5th harmonic was eliminated.

HARMONICS AND POWER FACTOR

As described in Chapter 2, the angle between kW and kVA defines the PF of the system in question. The larger the angle, the poorer the PF. Conversely, when the angle is very small, the PF is close to unity (1.0), resulting in an efficient handling of energy. The relationship for nonlinear loads is described graphically by the familiar PF triangle.

The only accurate way to measure nonlinear power factor is to measure the average instantaneous power, and divide that by the product of the true-rms voltage and the true-rms current. This is described graphically by the PF vector diagram described in Chapter 2 that includes a distortion vector.

Case history 1. An industrial customer asked us to size and apply a new unit substation. The reason? An existing 3000A bus system was carrying 3200A, and an additional 800A load was to be added. We asked about the system PF and were told that because there was no utility penalty for low PF, there was no need for improving it. The thinking was: no payback would be gained to offset the cost of installed capacitors. Nevertheless, we insisted on knowing what the PF was and asked for a copy of the latest power bill. It indicated a PF of a mere 0.58.

With this information, we explained this was the prime reason the total amount of amperes on the bus was so high. The system was being required to carry a large portion of nonworking energy (reactive amperes) just to get the relatively small amount of working energy.

Here, we wanted to reclaim from 1300A to 1400A from the system, thus returning the overall load, including the new load, to some level below the 3000A rating of the existing bus system. We noted that improving the PF to just above 0.70 would release 330VA (.33kVA) for every 1kW of load. This would net out to a 33% reduction in loading. As such, a 33% reduction of 4000A (3200A plus 800A additional) would result in a release of 1333A. Thus, the revised bus loading would be 2667A (4000A less 1333A).

The resultant ampere savings would come from providing vars locally. In this case, the capital cost savings were almost as dramatic: Capacitors were only 40% of the cost of a new substation.

Case history 2. A facility engineer attended a seminar and noted how nonlinear loads negatively affect PF. With this in mind, he went out to the various unit substations to read their respective PFs. At one location, he read a PF of 0.40! Based on this alarming information, he requested us to perform a PF and harmonic analysis.

Since new loads were anticipated, the capacity of the plant's existing substation came into question. Two loadcenter-type unit substations at this location were directly affected, and the solutions to their respective problems could not have been more opposite.

At the first loadcenter, which was where the 0.40 PF reading was taken, we found the engineer's PF meter was, in fact, out of calibration. The actual PF was 0.82. Thus the facility engineer knew it was reasonably safe to permit additional loading on this loadcenter.

At a second loadcenter, however, there was cause for concern. A large, 6-pulse DC converter was powered from the loadcenter's secondary bus. Due to the converter's demands for 5th and 7th harmonic currents, this bus was being overloaded. Also, when the converter was off line, the loadcenter's PF was good; when it was on line, the PF was very low. This is a classic tell-tale sign of nonlinear load demand.

Our recommendation was to install a harmonic trap sized for the rating of the DC converter and applied at its input terminals.

TRIPLEN HARMONICS AND NEUTRAL OVERLOADING

In a balanced 3-phase system, neutral current consists of the imbalance of the phase conductor currents (Ia, Ib, and Ic). As explained in Chapter 2, if these currents consist only of the fundamental (60 Hz) current, there is a 120° phase shift between each phase and the summation of these currents at every instant in time is zero.

If the phase conductor currents contain both fundamental and odd triplen harmonic currents, the result is very different. Triplen currents will add in the neutral conductor.

Case history 1. A facility manager of a data center was informed that additional file servers would be installed and would be powered from an existing panelboard. Therefore, an analysis of the panelboard, its feeder, and loading needed to be done.

To find out if a problem would occur with the proposed load addition, the facility manager asked the maintenance electrician to review the feeder of the existing panelboard, which consisted of three 500 kcmil phase conductors, a No. 1/0 grounded conductor, and a No. 1/0 insulated grounding conductor. All conductors had 75°C insulation. Testing indicated loads of 99A, 130A, and 77A respectively for the phase conductors and 130A for the neutral conductor. The load on the feeder after the addition of the new file server was estimated to be 132A, 182A, and 101A respectively for the phase conductors and 197A for the neutral conductor.

The electrician noted that an overload problem would occur, not with the phase conductors, but with the neutral conductor. The phase conductors would be loaded to approximately 60% of their ampacity for the worst case (B phase) while the neutral would be overloaded to 131% of its ampacity.

We saw the problem was more than just an imbalance on the phase conductors. Using the equation:

$$I_N = \sqrt{(I_A^2 + I_B^2 + I_C^2) - (I_A I_B + I_B I_C + I_C I_A)}$$

a value of 46A was obtained for the present load and 71A for the estimated revised load. The measured neutral current of 130A was obviously more than 46A calculated.

The cause of this problem was harmonics. The load on the existing panelboard was line-to-neutral connected nonlinear loads. This type of load consists primarily of odd triplen harmonic currents (3rd, 9th, 15th, 21st, etc.). These currents add in the neutral conductor and can cause measured neutral currents as much as twice the phase conductor currents.

In relation to the potentially overloaded neutral conductor, three options were developed to address this problem.

• Replace the feeder with one that has a full-size neutral. This option was not considered due to the problems and time involved in replacing the feeder in this particular application.

• Use a neutral filter to reduce the anticipated neutral current within the rating of the conductor. This option was seriously considered. A neutral filter uses a zig-zag transformer that reduces odd triplen harmonic currents by a factor of 7.54 or greater. It uses the phase shift characteristics of a transformer to cause cancellation of the odd triplen harmonics. This would reduce the anticipated neutral current to a value less than the current rating of the conductor. The filter was reasonably priced and could be installed with little downtime of the feeder. The disadvantage of the filter was that it had to be installed adjacent to the affected panelboard. This took up floor space and meant additional cooling load to the center's cooling system.

• Relabel the insulated grounding conductor and reconnect it as a parallel grounded conductor. This proved to be the winner. The insulated grounding conductor (same size as the neutral conductor) was never used as intended. Other feeders to the data center did not have grounding conductors; instead, the metal raceway system was used as a ground. After discussions with the electrical inspector and facility manager, the decision was made to relabel the insulated grounding conductor, disconnect it, and reconnect it as a parallel grounded conductor. This doubled the capacity of the neutral. The consensus was that this situation was no worse than other feeders in the system. The anticipated neutral current would now be approximately 65% of the capacity of the conductor.

Case history 2. One interesting situation involved an electrical contractor who was not aware of the problem of harmonic interaction. We were called to an office facility, along with this electrical contractor, to investigate distribution system problems. While measuring the current on each phase of a 225A, 3-phase, 4-wire, panelboard feeder with a true-rms sensing meter, we observed good current balance, with readings in the range of 100A each. Before measuring the neutral current, we asked the contractor what he thought the readings would be on the neutral. Noting that the loads appeared to be balanced on the 3-phase system, he suggested that the neutral current would be very low, perhaps only 3 or 5A.

When we closed the current clamp onto the neutral, the reading was 195A. Upon seeing this, the contractor exclaimed, "Your meter is broke!" "What about the phase readings?" we asked. The contractor replied they were correct but not for the neutral reading. His reasoning was that all 60 Hz currents balanced on the phase wires will cancel on the neutral. And he was right. What we were measuring on the neutral, however, was the third, ninth, fifteenth, etc. harmonic currents and, as the contractor originally expected, very little 60 Hz current. Obviously, he was unaware and certainly not knowledgeable of the harmonic interaction problem.

CABLE DERATING AND NONLINEAR LOAD PANELBOARDS

Installing a nonlinear load panelboard triggers certain NEC requirements that can significantly affect the size and ampacity of feeder wire or cable. In Note 10(c) to the Ampacity Tables of Sec. 310, the NEC requires that on a 3-phase, 4-wire wye circuit where the major portion of the load consists of nonlinear loads such as electric-discharge lighting, electronic computer/data processing, or similar equipment, harmonic currents are present in the neutral conductor, and the neutral must be considered a current-carrying conductor.

A 3-phase, 4-wire feeder to a nonlinear load panelboard will have, therefore, either four or five current-carrying conductors, depending on the feeder configuration. Note: Some makes of nonlinear load panelboards have three full-size neutral lugs. As such, three full-sized neutral conductors are possible in the feeder, resulting in six current carrying conductors.

In the past, Note 10(c) has often not been enforced because of the difficulty in establishing that a major portion of the load consists of nonlinear loads. With the installation of a nonlinear load panelboard, however, this difficulty is eliminated and inspectors can enforce NEC conductor ampacity derating requirements. In this situation, conductor ampacity derating will be per Note 8(a) and its accompanying table. Typically, this will result in a feeder conductor or cable being derated to 80% of its Table 310-16 ampacity.

Case history. We had specified a standard, 100A, 3-phase, 4-wire, main-lug-only panelboard for an unfinished light commercial space. For cost savings, we specified the metallic feeder conduit is to be used as the ground return path. Because we had no idea of the amount of load unbalance, we specified a full size neutral conductor. Standard main lug configurations can accept No. 10 through No. 2/0 wire. With this information, we specified a feeder of four No. 3 copper THHN 90°C conductors in an 1¼-in. EMT conduit, noting that the percent voltage drop was within NEC limits.

Everything seemed fine because we had been told the load was anticipated to be linear. As such, the feeder conduit contained only three current-carrying conductors and no derating was required.

Before construction began, along came the new tenant: (a word processing firm using many PCs) that also requested new fluorescent lighting with electronic ballasts. We reevaluated our original design and replaced the standard panelboard with a nonlinear load panelboard.

Obviously, with a major portion of this proposed load being nonlinear, the feeder neutral sizing had to be increased to handle 3rd-harmonic currents. For the new neutral conductor, we first looked at two No. 3 THHN 90°C copper conductors. Per Note 10(c), we would have five current-carrying conductors. Per Note 8(a), these conductors must be derated to 80% of their Table 310-16 ampacity, or to 88A.

Since this ampacity is below what was originally specified, we had to look at the next larger sized conductor whose derated ampacity is 100A or more. We developed the table shown in **Fig. 11.6**. No. 2, 90°C copper has an ampacity of 130A. Thus its derated ampacity is: 130 × 0.80=104A. The nonlinear load panelboard's main lugs will accommodate this size conductor. Therefore, we specified a new feeder configuration of five No. 2 AWG copper THHN 90°C conductors. Per NEC Table 3A, these conductors will fit in the 1¼-in. conduit.

The choice of feeder conductor sizing was verified by the fact that: the feeder overcurrent protective device would accommodate the continuous portion of the assumed load and protect the feeder conductors; the conductors would not overheat under the assumed loading and conditions of use; and the calculated load would not overheat the nonlinear load panelboard main and neutral lugs.

COMPUTER SIGNAL-REFERENCE GROUND

Electrical engineers have analyzed why the 60-Hz (power) grounding system is incapable of conducting RF signals to the common reference point (grounding electrode system), thereby equalizing the differences in potential. A properly designed power grounding system is of sufficiently low-impedance at 60 Hz to equalize any potential differences so that enclosures, raceways, and all grounded metal are at the same reference (ground) potential. However, at higher frequencies, equalization is prevented because of increased impedance.

The impedance of a conductor consists of three basic components: resistance, capacitive reactance, and inductive reactance. Although the inductance (L), in Henrys, will be constant for a conductor of a given length and cross-sectional area, the inductive reactance (X_L), will vary according to the frequency (f) of the applied voltage as follows:

$$X_L = 2\pi fL.$$

			Conductor Derating Table					
Row No.	Interior type	Lug wire range	Standard conductor size	Standard conductor ampacity (90°)	Number of current-carrying conductors	Ampacity after derating to 80%	Rerated conductor size	Rerated conductor ampacity at 80%
1	100A MLO	No. 10-2/0 Cu	No. 3 Cu	110A	5(ABCNN)	88A	2 Cu (130A)	104A
2	100A MLO	No. 10-2/0 Al	2 Al	100A	5(ABCNN)	80A	1/0 Al (135A)	108A
3	125A MLO	No. 10-2/0 Cu	2 Cu	130A	5(ABCNN)	104A	1/0 Cu (170A)	136A
4	125A MLO	No. 10-2/0 Al	1/0 Al	135A	5(ABCNN)	108A	3/0 Al (175A)	140A
5	225A MLO	No. 6-300 Cu	3/0 Cu	225A	5(ABCNN)	180A	250 Cu (290A)	232A
6	225A MLO	No. 6-300 Al	250 Al	230A	5(ABCNN)	184A	350Al (280A)	224A

Fig. 11.6. Table of conductor derating lists the respective normal and derated feeder conductor sizes for various sizes of nonlinear panelboards.

At 60 Hz, $X_L = 2(3.14)(60)L$
$= 377\,L$.
At 30 MHz, $X_L = 2(3.14)(30 \times 10^6)L$
$= 188.5 \times 10^6 L$.

By dividing the value of inductive reactance at 30 MHz by the inductive reactance at 60 Hz, it can be seen that with an applied voltage at a frequency of 30 MHz, the "equivalent" resistive value of a given conductor will be 500,000 times as great as the same conductor when the applied voltage is at 60 Hz. At 60 Hz, the equipment grounding conductor is a short circuit, and at 30 MHz, it is effectively an open circuit.

In addition to an increased inductive reactance at higher frequencies, there is stray capacitance and inductance between adjacent conductors or between conductors and adjacent grounded metal, as well as resonance effects. These also contribute to an increase of apparent conductor impedance.

This points out the need for developing a grounding system capable of providing a low-impedance path for all frequencies from 60 Hz or less up to the RF level to ensure that all connected equipment and enclosures are at an equal potential. Thus, proper grounding for computers and other equipment containing solid-state devices requires an equipment ground and an SRG.

Transmission line theory on the effects of high-frequencies says that standing waves will not cause a significant voltage difference between two ends of a conductor if the length of the conductor is not more than 1/10 to 1/20 of a wavelength. The wavelength of a 10-MHz signal is about 100 ft, and a 1/20 wavelength is about 5 ft. At 30 MHz, the wavelength is about 32 ft, and a 1/20 wavelength is a bit under 2 ft.

If a number of conductors are connected in the form of a grid to create a multitude of low-impedance parallel paths, there should be a negligible difference of potential between any two points on the grid for all frequencies from 60 Hz up to the frequency where the length of one side of one square represents about 1/20 wavelength. Therefore, a grid made up of 2-ft squares should provide an effective equipotential reference (ground) between any two points on the grid for signals up to 30 MHz.

Case history. Although it is possible to construct an SRG in the field by interconnecting a number of bare copper conductors, specifications on this project called for the use of a prefabricated grid. The SRG specified consists of 2-in.-wide, No. 26 flat copper strips arranged to form 2-ft squares. Interconnection of the intersecting strips was done at the factory using exothermic welds on 2-ft centers. The welded connection provides noise-free integrity that cannot be compromised as easily as a mechanical connection because the molecular bond of a welded joint is not affected by dirt, corrosion, cleaning fluids, or normal equipment vibration.

Connection between the solid-state equipment enclosures and the SRG was provided for by $3/32 \times 1.2 \times 36$-in. braided copper straps. A mechanical connection at the equipment enclosure and a field-executed exothermic weld secured this strap to the SRG.

A transient suppression plate (a 4×4-ft, No. 26 sheet of copper) was used underneath panelboards and power junction boxes located beneath the raised floor to prevent any noise currents present in the concrete slab reinforcing bars from being coupled into underfloor data cables. Additionally, every sixth metallic raised floor pedestal was bonded to the SRG. Again, field-executed exothermic welds at both ends of the copper bonding conductors connected the pedestals and the SRG.

The prefabricated SRG was relatively straight forward to install since this was new construction. The prefabricated grid was delivered to the jobsite in rolls that were 12-ft wide and up to 100-ft long. Installation involved rolling the grid out onto the bare concrete floor starting flush against one perimeter wall and moving towards the other.

As obstacles were encountered, such as steel building columns, the grid was cut, fitted around the base of the columns, and welded back together to form, as closely as possible, the desired 2-ft squares. Steel building columns were bonded to the grid by 2-ft lengths of No. 6 bare copper bonding conductors. In cases where the column was larger than the 2-ft square, the size of the square had to be increased, but not more than absolutely necessary because the 2-ft square sizing of the grid is based on the fact that at 30 MHz, a 1/20 wavelength is a bit under 2 ft. The greater the deviation, the greater the effect on the SRG's capacity to perform its desired function. Where one roll was laid next to another, the outer edges of the rolls were overlapped to form a single strip and then exothermically welded together at 2-ft intervals.

To achieve the best performance from the SRG in terms of eliminating high-frequency disturbances, all metallic objects under the raised floor (conduits, water piping, ducts, steel building columns, etc.) were bonded to the SRG. As the various underfloor piping and conduit systems were installed, each in turn was connected to the grid by a 7-strand No. 6 bare bonding conductor. Additionally, bare bonding conductors were used to "jumper" all connections and fittings along the length of these piping systems to ensure a sufficiently low-impedance path along the entire run. The shortest possible length of bonding conductor

Prefabricated grounding grid was delivered to the jobsite in rolls. The grid was secured at one end of the building and then rolled out on the floor slab.

When building columns or other obstructions were encountered, the prefabricated grid strap was cut and then thermally rewelded together to minimize distrubing the 2×2 ft. pattern.

was used so that the desired maximum of 2 ft was not exceeded except where absolutely necessary.

GROUNDING INSTRUMENTATION AND CONTROLS

Process-control computers, instrumentation, and programmable logic controllers (PLCs) operating at 5V have replaced 120V relays, timers, and sequencers. Whereas slow, bulky, low-impedance electromechanical devices were almost invulnerable to electrical interference, the new generation of microprocessor-based control and monitoring devices is very susceptible. Installation techniques that are not tailored to these devices can lead to unexpected halts and faults, execution and reading errors, unpredictable operation, and deadly intermittents that are agonizingly hard to trace.

Knowledge of logical steps to take and the reasoning behind them will allow the design of a facility that is more likely to be trouble-free. Such knowledge is also critical before undertaking actions to help an existing facility having interference problems.

Limitations of NE Code grounds. Electrical grounding required by the code is not effective for control-system signal grounds. Its purpose is to provide an assured path for fault current to reliably trip the circuit breaker and remove power. Any signal grounding included in this equipment grounding is not intentional, just a bonus.

Differences between power frequencies and control system frequencies are the reason the NE Code grounds are inadequate. As frequencies rise, ground wires become antennas and progressively suffer from skin effect. An alternating current at only 7.5 kHz (which can be heard on a loudspeaker) will penetrate the equivalent of only 32 mils (.032 inches) in copper, which is one of the best conductors. At microprocessor switching speeds over 1 MHz, current penetration in copper is less than three mils (.003 inches). The result of skin effect on effective impedance between one point and another is pronounced.

Of equal importance as frequency rises is the fact that wavelength proportionally decreases. A quarter wavelength at 10 MHz is just 24 ft in free air and less when flowing on the surface of a conductor. Circulating currents see an apparent open circuit at one-quarter wave intervals, so the current path is unreliable. Some frequencies pass and some do not, causing distortions. This is why short returns are always recommended for fast rise-time circuits, and long, small diameter return wires (such as NE Code ground wires) act more like antennas and chokes than signal returns at system signal frequencies.

Any current given two paths will proportionally take the path of least resistance. If one path is 10 ohms and the other is 1000, the split will be 100:1. If one path is 10 ohms and the other is 0.1 ohm, the split will be 1:100. Ten ohms is low impedance in the first case, and high impedance in the second. A microprocessor at the end of that 10-ohm path would have to cope with possibly destructive currents in the first case and would be protected from them in the second.

In a system that is tied together, all impedances cannot be high. Thus, it is necessary to provide low impedance where current should flow or where low voltage drop is desired, and raise impedance only where isolation is required. The contrast created between the strategically placed high- and low-impedance paths will function as an effective distributed filter.

The places where currents should flow or where voltage drops should be kept low must be located by a detailed tracing of where circuits go and how they return. The low-inductance paths must then be installed to provide a clearly identified, single low-impedance path to which an equipment chassis can be tied.

To make an entire facility act as a distributed filter, use the required NE Code grounds, low-inductance copper signal return paths, isolation RF chokes, Faraday-shield transformers, powerline filters, metallic conduits, and other common devices. Such a system will meet all safety requirements and provide inherent isolation of sensitive equipment from electrical noise.

Fig. 11.7 shows how ground connections should be made for a process containing microprocessor-based controls, system power sources, and system utilities in order to reduce ground loops. Process utilities such as blowers or pumps should not share return currents with a process controller. As shown, separate ground returns should be provided for each. The ground return for each part should be copper and have the maximum area. A 12-inch-wide copper sheet or 4-inch-diameter copper pipe (both at least 1/16-inch thickness)

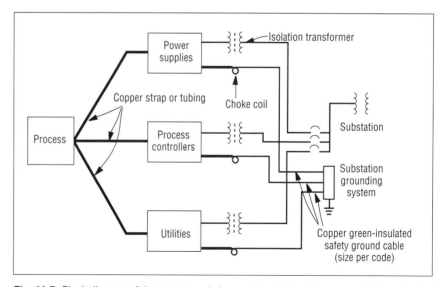

Fig. 11.7. Block diagram of the recommended grounding of a typical process that is controlled by microprocessor equipment. Each major component has an independent low-impedance signal return to the process center. Return currents to noncontrol equipment are separated from return currents to controllers, so ground loops do not exist. Interaction is limited because of the shielded transformers and RF chokes.

would not be overkill for each return.

A single-Faraday-shield transformer with a series capacitance of 30 picofarads (instead of the 2000 picofarads with a standard transformer) should feed each major system element. These single-shielded transformers are readily available.

The major conceptual stumbling block is the legal re-derivation of the transformer's downstream ground. If the incoming and outgoing grounds are not connected together, the path back to the breaker mandated by the NE Code has been broken. On the other hand, if they are connected together, the interactive paths are reestablished.

There is, however, another alternative: use an RF choke to continue the ground path. A choke will ensure that higher-frequency signals do not use safety ground leads as a way to get around. Also, the code requirement for continuous conductors of sufficiently low impedance to trip the 60-Hz circuit breaker is complied with while reducing the potential for bothersome loops.

Twelve turns of tightly wound green-insulated wire on a 6-inch diameter yield an RF choke of about 50 microhenries. When installed as shown in **Fig. 11.8**, the choke becomes part of the continuous safety ground, but electrically separates the system at signal frequencies. For instance, at 60 Hz, the choke looks like 0.02 ohm (which will easily pass breaker trip currents), while at 7.5 kHz, it presents a 2-ohm series impedance back toward the substation from the transformer secondary. Thus, the ground path toward the process, where the heavy ground is installed, will look a hundred times lower.

Equipment rack grounding. In looking for control system signal returns, be wary of the ability of rack systems, conduits, and wireway to provide the needed signal return path. Many racks are made of steel and thus present significant resistance to current flow. This resistance is likely to be seven or eight times higher than copper at DC and lower AC frequencies, and gets worse because the ferromagnetic character of steel causes higher impedances at higher frequencies due to skin effect. In addition, painted and bolted joints that increase resistance are often encountered in racks.

Equipment racks in each major area should be tied together with copper strips placed either on top or bottom of the equipment, serving as an extension of the grounds running to the process. This means that everything in those racks, including all instruments and controls, have a direct low-impedance connection to the process.

Such an approach seems to be defeated when racks are installed on concrete floors. Conventional wisdom says that the concrete also is a conductor. However, this is not an important consideration because when a .01-ohm signal return has been provided, the 2000- to 5000-ohm conduction path of the concrete will be swamped out. Thus, the racks can safely be bolted directly to the concrete.

The equipment case of all microprocessor-based equipment should be connected with a strip to the rack copper strip. Each rack plugstrip used to power the units should be fed through a powerline filter. This will shunt the transverse-mode noise the isolation transformer cannot effectively exclude back toward the transformer via the system copper strip.

Equipment ground wires installed in steel conduit will resist carrying signal return frequencies because of the increased RF impedance induced by the surrounding ferromagnetic material (the conduit).

HINTS ON INSTALLING ELECTRONIC CASH REGISTERS

For most electronic cash registers (ECRs), a dedicated branch circuit is an absolute requirement. For many, an insulated ground (IG) circuit is necessary to minimize the effects electrical noise can have on ECR operation.

IG circuits. There are several advantages to a dedicated IG circuit.

• The breaker protecting the circuit is less likely to trip. This is important, since a power disruption does more than interrupt an ongoing transaction. It can lock up the register, requiring use of a special service key to restore it to use; or, in the worst case, an outage can wipe out its programming.

• The hot and neutral wires are not serving other outlets that may have inductive or switching loads. Such loads can result in power surges, sags, and spikes.

• The grounding conductor is more likely to be free from ground-fault and harmonic currents originating from extraneous sources.

• Due to the manner in which these circuits are usually installed, all conductors have a degree of protection from induced voltages such as are caused by radio-frequency interference (RFI).

Raceway requirements. A dedicated IG circuit, at minimum, has three insulated conductors (hot, neutral, and ground) running from the outlet to its source of power. Although not required by the NE Code, ECR manufacturers recommend that these conductors be run in a metallic raceway or armored cable. The metal raceway or cable armor is grounded, through its connectors, to the panelboard at which it originates. The raceway serves to ground the metallic box in which the receptacles are mounted. Together, these provide a reasonable facsimile of a Faraday cage protecting the conductors of the IG circuit.

A solid raceway (rigid metal conduit, IMC, or EMT) rather than flexible conduit or armored cable is sometimes recommended for increased reliability. The use of flex may require the addition of a separate equipment grounding conductor for raceway and outlet-box grounding.

Use of extension cords as a substitute for permanent wiring to electronic

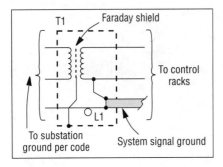

11.8. *Typical single Faraday-shield isolation transformer (T1). Choke coil (L1) is made up of 12 turns of insulated wire on a 6-in. diameter that have been tightly bundled together with cable ties.*

cash registers can be a violation of the NE Code and increases the chance of damage or disruption due to EMI. Cords do not have the shielding provided by a grounded metallic raceway.

The receptacle fed by an isolated-ground circuit has a triangular marking on its face. This is an industry standard adopted to comply with the provisions of the NE Code. The difference between it and a standard receptacle is that there is no electrical contact between its yoke and its green (grounding) terminal screw and associated "third-prong" contact.

When only one ECR is to be served, a single receptacle should be installed, not a duplex. There is a possibility that the second outlet of a duplex receptacle would be used for some item such as cleaning equipment or other heavy inductive load. Duplex receptacles can be used when multiple ECRs are powered off one IG circuit. It is good practice to provide nearby receptacles on non-IG circuits so that other equipment can be conveniently used nearby.

If more than one IG circuit is run, ECRs and related devices that intercommunicate should be on the same panel phase-leg. A word of caution: some communication-type cable has a pull strength of only 18 lbs, so it cannot be handled as though it were armored cable during installation.

Other IG circuit tips. ECR manufacturers differ on where the IG circuit should originate. One, for example, insists that the ECR circuit should come from a separate subpanel rather than share a panel with non-ECR circuitry.

One agreed-on rule is that power should not be taken off a phase bus that serves heavy inductive loads. This is not always possible. However, if one phase is heavily loaded with fluorescent lighting while another handles mostly show-window incandescent spotlights, the latter is preferable for ECR power.

The insulated grounding conductor of an IG should be run through a panel or subpanel without connection, and be terminated at the ground/neutral bus of the main service panel.

Wiring of IG receptacles is also different from the way normal receptacles are wired. **Fig. 11.9** shows the correct wiring method to use at an isolated-ground receptacle.

Installation pitfalls. The source can be tested for inherent noise with a recording oscilloscope or peak voltmeter. Some clamp-on meters have a peak setting that will retain spike readings, but that means leaving it in the panelboard for a number of days while determining whether an IG line should be installed. A further problem is that a peak recorder does not show sags. Voltage sags may result in damage equal to or greater than that caused by spikes. Also, damaging spikes may be of too brief duration to register on a meter not specifically designed for the application.

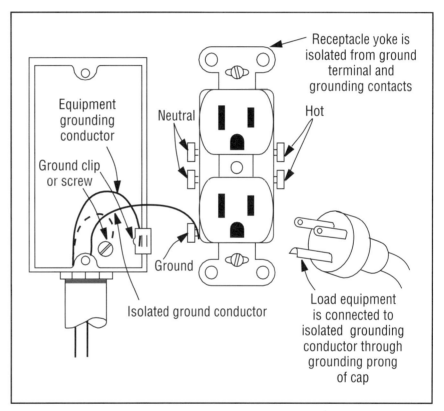

Fig. 11.9. *Grounding of an isolated-ground receptacle. An IG receptacle grounding terminal is connected directly to the isolated ground conductor. Box and isolated yoke of receptacle are grounded to building ground (if necessary) by a separate equipment grounding conductor.*

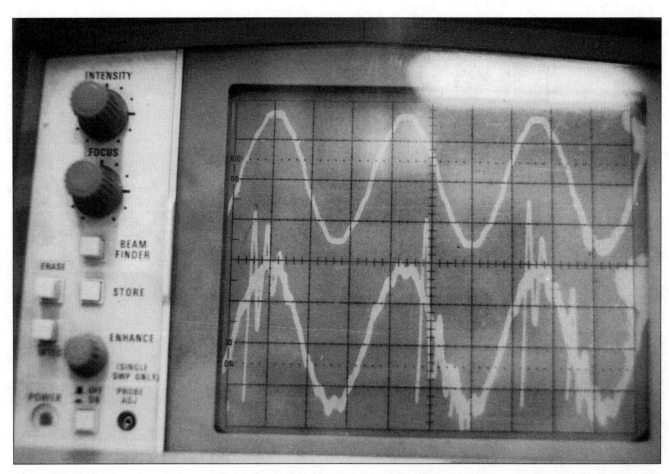

Diagnostic instruments that capture the nature of sinewave distortion are helpful in final selection of the type of power conditioning equipment that is required for the installation.

EQUIPMENT CASE HISTORIES

Specialized equipment often needs to be applied to a power distribution system in order for solid-state electronic equipment it feeds to operate satisfactorily. Here are some actual case histories where this type of approach has been used. It will provide a better understanding of the reasoning behind the application of some of these items.

HARMONIC FILTERS

The key to eliminating harmonic problems is the use of filters at the respective nonlinear load. Since a filter is designed to reduce the amplitude of one or more fixed frequency currents, a harmonic filter is one means of reducing waveform distortion. Since the most important harmonic frequencies are very close to the fundamental power frequency (60 Hz), a very selective filter is required. This specially tuned filter will tune out the dominant harmonic being generated.

A frequency selective harmonic filter can be used as a series filter, as was described in Chapter 3 of this book. A series filter uses a high impedance to block the power source from the flow of harmonic current being generated by the nonlinear load.

A shunt filter, on the other hand, provides a low impedance path so that the harmonic current will divert to ground.

Although either method of filtering will reduce harmonics, the shunt filter is usually preferred, since the series filter must be designed for full line current and insulated for line voltage.

Case history. Spectrum analysis data of a 12-pulse, 200kVA uninterruptible power supply (UPS) module installed without input filters was collected in a computer data facility. Readings were taken of Phase-A-to-neutral voltage and Phase-A current. The data obtained represents the distortion caused by the UPS on the input voltage and current. The

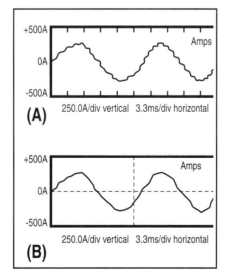

Fig. 12.1. Current waveform before harmonic filter was installed (A); and following the installation (B).

THD of Phase-A-to-neutral voltage was 5.5% of the fundamental 60-Hz voltage, while the THD of Phase-A current was 16.2% of the 60-Hz current.

Filters were added to this equipment and spectrum analysis data was again taken. The THD of Phase-A-to-neutral voltage was reduced to 2.7% of the fundamental while the THD of Phase-A current was reduced to 6.3% of the fundamental. **Figs. 12.1(A)** and **12.1(B)** show the current waveforms before and after the filters were installed.

Several UPS modules were retrofitted with input filters with similar results for each. The cost of the retrofitting filters was approximately $30 per kVA of UPS. For new UPS installations with manufacturer provided filters, the filter cost is estimated to be $10 to $15 per UPS kVA.

Our recommendation for most applications is to filter the harmonics at their sources (at the nonlinear loads). By filtering at the source, the magnitude of harmonics flowing on the electric system will be minimized. One approach that should be looked at is to have the manufacturer provide filters that limit their equipment's current distortion.

MG SETS

One of the real concerns for operating engineers is in the maintenance area. What can be expected in cost and other troubles possibly awaiting the user of an MG set? In our experience over the years, we find the maintainability of this technology far more user friendly than other technologies. For example, most solid-state products usually require service from the original manufacturer, while many of the motor/alternator products can be serviced by local staff maintenance personnel or a local motor shop with a relatively small amount of specialized parts. Conversely, the static UPS usually has very distinctly designed parts for each unit, and the proviso that factory service should be used rather than third-party offerings. Electronic parts upgrading takes place frequently, and the changing of components over the life of a unit has been a problem in the past. In the case of the static units, a factory service policy complete with a spare parts kit is recommended to be purchased with the original order placement.

Case history 1. On one occasion, where a motor-alternator was installed on the top floor of a facility, our client experienced a premature winding failure in the motor, certainly the worst of all possible failures in the worst of all possible locations. Yet, when the service man from the local motor repair shop came to the site, he gave an estimate of three days for a total disassembly of the unit. This meant taking the parts out piecemeal, completely rewinding and rebearing, retesting the component parts, and returning them to make an operating system. We were challenged to think of how to accomplish such an overhaul for a static system. Just as he promised, all went back together in three days, and this unit has continued to perform as specified over the ensuing seven years.

One might look at this technology for its ability to assist in swallowing up harmonic currents coming from the devices on the output circuits. While the alternator is not designed as a harmonic trap, it does manage to serve like any other coil inductance. When the output impedance remains low, and the harmonic current loading is a small percent of the total current, the unit can perform well in handling the harmonic current without significant voltage distortion and without allowing transfer of the harmonics to the incoming circuit.

A bonus for the facility power factor improvement can be achieved when using a synchronous motor as the input driver. By adjusting the field control on the synchronous motor, it can be changed from lagging to leading power factor, now supplying vars (volt-ampere reactive energy) to the system at no additional expense.

This method of assisting power factor improvement may be one of the strongest available, and one not subject to the possible damaging effects that harm capacitors. Maintainability and reparability seem to give the concept the possibility of a long life. The ruggedness of construction combined with the additional features listed above may make this a good tool, especially in harsh factory conditions.

Case history 2. We have seen instances where MG set performance traits has provided a high level of protection in many lightning-prone areas of the country. Probably the most significant of these instances was in the Southeast portion of the U.S., where a company, in the past, had experienced severe damage from lightning. The repeated "ON-OFF-ON-OFF" cycling of the power during these lightning strikes had caused so much damage to computers and resulted in so much cost for equipment replacement and repair labor that the facility's computer service company canceled its annual agreement.

We were called in to recommend a solution to the problem. Our suggestion: A synduction MG set. After the installation, several pieces of computer equipment were allowed to remain powered by unconditioned utility power, while the balance was conditioned through the synduction-motor-driven MG set.

The next storm arrived, and all the equipment running on the MG set stayed up and suffered no damage. However, the equipment on unconditioned utility power was extensively damaged. Here was a very dramatic, side-by-side comparison of the protecting power of the synduction MG set.

Case history 3. A railroad switching center in the southeast regularly used a small computer to count freight cars brought into its switch yard and to dispatch trains as they went out of the yard. The owner requested help after receiving a notification from its CPU vendor stating its maintenance contract on the equipment would be terminated. It seems that the cost accrued on the contract in the course of one year's maintenance was over $26,000. Thus the vendor decided to terminate the agreement because it was installed without any power conditioning protection.

We were asked for a recommendation of the appropriate power protection equipment. In light of historical storm and power company switching events, we recommended a motor generator (MG) set. The overall pricing of the equipment was in line with the owner's budget and the recommendation was accepted.

We then were invited to train those who would be operating this equipment and began by explaining how the MG set would work and what type of protection it would provide. Our audience expressed disbelief that such an installation without battery backup would work. This disbelief was to become the basis for a very interesting revision in the installation after we left the site. It would also would provide an excellent test of the protection offered.

After final testing and certification of the MG set, we turned the project over to the owner, with the installation as shown in **Fig. 12.2(A).** Before there was an opportunity for the installed system to prove itself, the facility engineer decided not to put all his eggs in one basket. He removed the peripherals from the MG set output and rewired them to the regular electrical distribution, as shown in **Fig. 12.2(B).** This, he thought, would preclude an MG set performance failure from affecting both the CPU and the peripherals. He still did not believe in the capabilities of the MG set.

Some 8 to 10 weeks later, a severe thunderstorm with extensive lightning activity developed and lingered in the area for almost a week. During the storm, the utility's lines dropped to 0V for fractions of a second, one phase at a time. Switching of supplied power from alternate utilities to the grid occurred frequently and continued throughout the week causing considerable problems to the entire area affected by the storm.

We went back to see how the installation had worked throughout the storm and asked the facility engineer. His response was, "You made a believer out of me!" At that point, we still were not aware of his revision to our installation. He then told us about his revision, which had resulted in another $5000 to $6000 worth of damaged because the storm had created such havoc with the voltage on the main distribution of his facility. By contrast, the MG set had served as a buffering unit by keeping all the discontinuities generated during the storm from affecting the CPU unit, which was still on the MG set output. It was a hard lesson to learn. Thereafter, the MG set was treated with a great deal more respect.

We were thankful for the opportunity, unknowingly provided by the facility engineer, to conduct this side-by-side test. The end results reinforced the application recommendations. Here we see the beauty of an electrically separate input and output, along with the benefits of the mechanical connection between the motor and the generator. Over the years we have been able to observe that the application of MG sets provide a high level of protection against lightning.

SUBCYCLE STATIC TRANSFER SWITCHES

As was discussed in Chapter 5, not all solutions to quality power shortcomings have to depend upon the installation of power conditioning equipment. Sometimes the answer is as simple the addition of high-speed transfer switches that will provide a means of securing a continual source of power. Of course, two independently derived sources of power are required for this solution to be applied.

Case history 1. We were asked to come to a bank's operations center to recommend specifications for a new static UPS that was to serve a critical computer room. The owner showed us the existing electrical service, and we obtained transformer size information so we could do the required load calculations for sizing the proposed UPS. We immediately noticed there were four separate power lines entering the facility, along with a standby turbine generator for backup.

We asked the owner why such a service arrangement needed a UPS in addition to the multiple power sources; high-speed transfer switch technology could do the job just as reliably yet much less expensive. The owner had never heard of such technology.

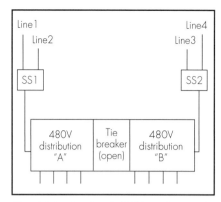

Fig. 12.3. A pair of 480V static transfer switches, one for each pair of power lines feeds a double-ended substation. In this configuration, transfer switching is fast enough between the multiple utility lines that any single-line disturbance will not cause a shutdown.

One of our proposals for the site included installation of a pair of low-voltage static transfer switches, one for each pair of power lines. Both switches feed a double-ended substation (See **Fig. 12.3**). In this configuration, the power sources can be switched fast enough between the multiple utility lines that any single-line disturbance will not cause a shutdown. Thus, the redundancy of the service design is preserved.

Case history 2. Here is a look at a few typical customer power applications, concentrating on the 1.5kV and 25kV distribution voltage levels. MV subcycle static transfer switch applications are derived from field studies made at various customer locations. They demonstrate the diversity of available solutions, even when the technology seems to be limited to a narrow range of sites, namely those having two independent power sources.

<u>Application No. 1</u>. A semiconductor manufacturing facility with 14,000kVA of load was faced with a UPS budget of more than $4 million. Even when all the sensitive loads were concentrated, the same amount of battery support protection was needed. Since the facility had the required two independent power sources, the company was able to reduce the cost expectations, through the use of a subcycle MV transfer switch, down to below $600,000. Yet it achieved similar protection.

<u>Application No. 2</u>. An industrial fa-

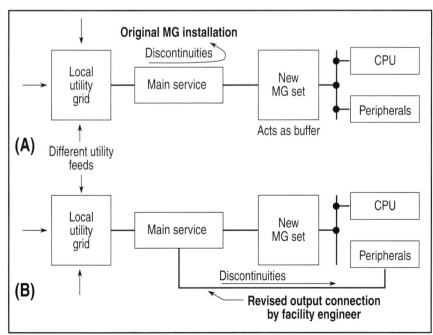

Fig. 12.2. MG set connection as turned over to owner (A); and as it was rewired (B).

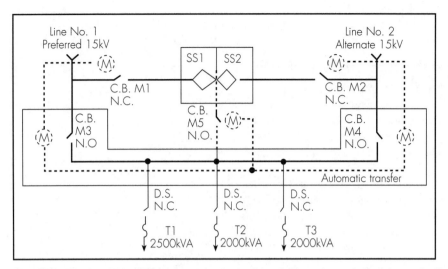

Fig. 12.4. An industrial facility had an automatic transfer switching scheme that used motorized circuit breakers (M3 and M4). The recommended solution was to use MV high-speed transfer switches (SS1 and SS2) as the normal circuit between Lines 1 and 2.

cility was facing yearly losses of $700,000 from process downtime due simply to circuit breaker operations. As shown in **Fig. 12.4**, the company already owned an automatic transfer switching scheme of the "slow" variety (shown with the dotted line) that used motorized circuit breakers (designated "M3" and "M4"). The recommended solution was to use "silicon" technology transfer switches (designated "SS1" and "SS2") as the normal (high-speed transfer) circuit between Lines 1 and 2. The existing motorized breakers would serve as the backup and bypass. A one-year payback is estimated for this installation.

Case history 3. In some instances, it is better to first secure the alternate feed sources and let the decision to install subcycle transfer switches until after the customer feels comfortable relying on two alternate power sources. Such was the case at a major communications company. We learned that the firm had experienced many instances of downtime on its main switching equipment and computer operations as a result of power line disturbances. Some of these were very short time intervals, most being less than 1/2 sec. Some, however, lasted from a few sec to a few min.

In situations such as this, we always recommend the end user contact the utility, asking what kind of assistance could be expected. When we made this recommendation, the facility engineer's reply showed a total lack of confidence in the utility to help find a workable solution. "We use the kilowatts and they send us a bill; they certainly would not be of any help with this kind of problem!"

After hearing our reasoning for the suggestion, the engineer grudgingly gave permission to contact the utility. We asked the utility engineering contact person for the nearest location of a second power line or substation to the communications company facility. The person seemed baffled that a company would pay an additional facility charge when it already had electric service to its property.

We explained that the customer was faced with an $800,000 expense for an undesirable battery-supported UPS system, should an acceptable alternative not be found. The communications company only needed an alternate feed to its building at a "reasonable" facility charge. Several different techniques could then be used to ensure power continuity, saving the company 40% to 50% of the anticipated overall costs of a conventional UPS.

The utility engineering contact did the required research and found a second distribution substation across the street and behind a hillside from the communications company. The cost of a new underground feeder from this substation was quoted as $40,000 by the utility, considerably less than our anticipated $150,000. The facility engineer was pleased at the savings and learned that a utility can be part of a power quality solution; the utility learned how a search of its existing primary distribution network can save its customer considerable capital expense.

A new underground feeder was installed and the facility now had two 480V power sources for a dual feeder system. The facility engineer now had options to consider for distributing and conditioning power.

This electro-mechanical installation has served the facility engineer well and has saved capital expense while eliminating the anticipated maintenance headaches associated with an battery-powered UPS.

SURGE PROTECTION

All solid-state electronic equipment is sensitive to voltage transients due to lightning strikes. For that reason, some form of surge protection is required.

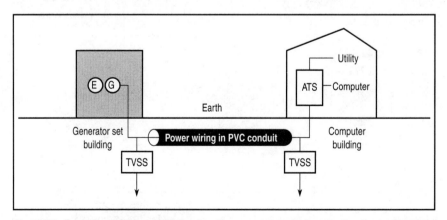

Fig. 12.5. Even though this installation has the proper transient protection for underground wiring between buildings, damage occurs to surge protective devices because the high generator-induced currents and oscillating voltages exceed the devices' performance level.

The fundamentals of the subject is covered in Chapter 7 of this book.

Case history 1. As shown in **Fig. 12.5**, an engine-generator set located in a remote building was providing standby power via underground wiring to a computer building containing the main electric utility service and automatic transfer switch (ATS). The installation followed recommended guidelines that protectors be installed at both earth entry/exit points of the standby power wiring (the remote engine generator set building and the computer building). These devices were providing phase-to-phase, phase-to-neutral, and neutral-to-ground protection. The 480V electric utility service was also protected where it entered the computer building.

As the installation was in the final stages of being completed and turned over to the owner, we received a call indicating that the transient protection devices on the standby power wiring had been damaged. We assumed that the damage resulted from a lightning strike that created larger surges than the ratings of those devices. We asked what type of a lightning storm occurred at the site and were surprised to hear that there had been no storm at all. We then asked what was taking place when the devices were damaged.

It appears that the engine generator was undergoing final testing when the voltage regulator on its output failed, causing the output voltage to oscillate wildly. In fact, it exceeded the voltage threshold of the protection devices and caused those devices to conduct to ground. Those who were in close proximity to these devices first smelled smoke, then saw smoke, and then heard the main circuit breaker trip to remove both the engine generator set and the faulted transient protectors from the line.

Here was a situation where the generator-supplied current, which was passing through the protective devices and going to ground, acted as if it was a short circuit. The transient protectors were not designed to sustain steady state currents of this magnitude; thus the elements inside were damaged.

One mystery involved the transient protector at the engine generator building. It sustained the least damage while the device protecting the standby power wiring to the ATS was almost completely destroyed.

This mystery was explained when we noted that the ATS had transferred to the "neutral" position and was no longer making the circuit continuous to the load. When a lightning charge is building up on an overhead power line, the resultant increasing energy acts as a traveling wave, traversing the overhead line. At any point where the wave encounters a "discontinuity", it then treats that point as a reflecting point. In reflecting back along the line, the traveling wave follows the laws of physics and doubles its voltage. If it were to travel into the reflecting point at 30kV, it would travel out of and away from the reflecting point at 60kV. This doubling effect takes place at dead-ends, open disconnect switches, or the front end of transformer primaries.

In this case, the open ATS was acting as a "reflection point" for the power wiring coming from the engine generator set. Thus, the voltage level produced by the engine generator and seen by the first protector was now doubled at the reflection point (the open ATS). As a result, the damage was much more extensive at the ATS's transient device.

The manufacturer compared this varying damage, indicated the specified performance levels of its devices, and confirmed the effect of this abnormal transient on the design of these devices. Everyone involved now understood that this was an unusual and unintended diversion of large fault currents through transient protective devices.

Future problems of this sort would be averted by using protective equipment with thermal cutouts that remove the device from the line before sustained damage can occur. This self protective feature operates when sustained current flow causes increased temperature inside the device.

Case history 2. The facility manager of a large data center complained about the large expense incurred due to frequent replacement of many driver and receiver boards in the facility. All the reported problems seemed to occur during lightning storms. Nevertheless, we asked if there were any other occasions of damage that were different from what had been described. We were told that one night all four of the boards had to be replaced three times as a result of three separate lightning strikes. At $800 per board, that storm alone caused $9600 of damage.

The facility layout is shown in **Fig. 12.6**, with two buildings in a campus-style layout connected by two 46 MHz RG62U data cables run in PVC conduit under the ground. An overhead electric power line runs across the back of the property, with a distribution lightning arrester located on a pole almost half way between the two buildings.

We asked why the RG62U cable was not shielded, since it was an outdoor installation. We were told this was not a true outdoor installation since the conduit was continuous from one building to the other!

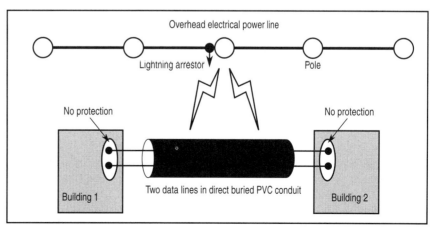

Fig. 12.6. *The function of the pole-mounted surge arrester caused the lightning-induced energy to flow through the earth to the unshielded, unprotected data lines. These, in turn, allow the transient earth currents to enter the data center buildings and damage electronic equipment.*

Our recommendation was the installation of a data-line protector at each end of each run, which would cost much less than one driver board. To apply the correct protection, the supplier of this equipment needs to know the operating voltage and frequency of the signal as well as the impedance match in the system. In this instance, the protected frequency range is 46 MHz, with an impedance requiring 50 ohm BNC connectors.

Case history 3. We were asked by the local electric utility to advise one of its customer on how to protect home furnishings from damage. Several lightning strikes had occurred in the past, destroying this customer's television set four times. After the last incidence, we viewed the damaged set and found a 4-in. hole in the side of its cabinet. We noted that duplex receptacles in the room were protected by plug-in transient protectors. We also noted what appeared to be a burned carpet area just beneath where the coax TV cable attached to the TV set. Upon closer inspection, however, we found the burned area to actually be carpeting covered with a thin coating of the coax cable's black polyethylene jacketing material. Where the jacket had melted off, the coax was silver colored.

Looking at the various services, we found that the electric, telephone, and cable TV systems each had their own service point into the residence: electric at the rear, telephone at the side, and cable TV at the front. Each of these services had no protection. In addition, there was no common ground connection to which all of the above services could be referenced.

Based on this observation, we theorized that a lightning-induced traveling energy wave came into the residence through the cable TV service, was looking for a path to ground, and found such a path in the power supply ground connection to the TV case. This created the hole in the TV cabinet and sufficient heat to melt the cable TV lead's jacket.

Our suggested solution was to bond the ground connection points of the three services together under the house in a crawl space. We also advised that a transient protector be placed at the cable TV service point. In effect, both

Power distribution units *provide a convenient place for final filtering out of noise on circuits supplying computers and other solid-state electronic devices.*

the electric and TV lines would be protected and referenced to the same ground potential. There have been no more calls for new TVs.

POWER DISTRIBUTION UNITS

A power distribution unit (PDU) is also often referred to as a computer power center (CPC) because it was often used as the distribution point for power delivered to data-processing computers and their peripherals. Now these units are used in association with varied types of solid-state electronic loads.

Case history. Installation of a 30kVA PDU in the computer room at a law firm assures clean power for a variety of computer-based electronic equipment. Solid-state electronic items include disk-drive storage units, laser printers, teletype printers, file controllers, and a local area network (LAN) controller. The LAN permits terminals located on five floors of the office building to exchange information with a centrally located computer. Telephone modems provide long-distance telephone line communications with the firm's branch offices in other cities. In addition, most personnel use PCs as part of their daily operations.

Since the computer equipment is used for the preparation and storage of briefs and correspondence, and for client billing, all of the firm's activities depend on having a highly reliable distribution and delivery system.

A power conditioning section and a power monitoring and distribution section are included in the PDU selected, and are contained within a single free-standing cabinet. The power conditioning section uses ferromagnetics (transformers and inductors) and capacitors to synthesize a high-quality 3-phase output that is free of transients, noise, and other disturbances. The input power is used only as a source of energy.

The magnetic synthesizer consists of six 3-phase pulse transformers connected in parallel with phase-shifting inductors, and a lineup of capacitors. It provides a 12-step low-distortion, 60-Hz, 3-phase sine wave that is similar to that derived from a UPS. The synthesizer does not contain any mechanically moving parts or power semiconductor components (except in the control circuitry). It offers high efficiency (94% for the 30kVA size at full load, with nearly unity power factor).

Since the output sinewave is electromagnetically generated, it is completely isolated and independent of all input parameters except for phase rotation and frequency. A short, an open, or an unbalance on any input phase has minimum effect on the system. And in the event that one incoming phase is lost, the 3-phase output is maintained (although the angle between phases will not be 120°) with only a small drop in output voltage, at loads up to 60% of full load. Even 100% unbalanced loading of phases does not cause a significant change in output voltage, and variations are held to less than 3%. Because of the unit's rugged construction, a short circuit up to 300% of the full-load current can be sustained without damage until overcurrent protection devices clear the fault.

The transformers and inductive winding array establish a single point ground that serves as the power grounding point and the computer system grounding point, minimizing common-mode disturbances.

A rigid metal conduit run beneath the 12-inch-high raised floor of the computer room carries 480V power to the PDU. The conduit is coupled to an underfloor junction box that has terminals for connecting three incoming feeder conductors and an equipment grounding conductor.

Output conductors from the PDU to equipment are enclosed in liquidtight flexible steel conduit with a specific length and a receptacle termination matching the connected equipment. Every output cable is permanently labeled with its circuit number and cable length at each end of the cable to assure positive identification when cables have to be rearranged. Also, new cables can be added and routed with minimum disruption.

The PDU's input molded-case main circuit breaker can be tripped manually, thermal magnetically to clear a fault, by a shunt trip activated by 180°C temperature sensors in the transformer windings, shunt-tripped by a local emergency power off switch, or by a manual-restart circuit. This circuit will shunt-trip the main input breaker whenever input power is lost. This prevents the PDU from being subjected to repetitive power turn-ons if the utility is involved in fault clearing operations.

From the synthesizer output, conditioned power at the utilization voltage (208/120V, 3-phase) passes to the distribution section. This output section consists of a main circuit breaker serving two vertical rows of branch circuit breakers. From the branch circuit breakers, power is delivered to the computer room equipment via flexible cables.

Located directly above the distribution section, a section monitors and displays output voltages/current/kVA, and audibly and visually indicates alarms for output overvoltage and undervoltage, transformer over-temperature, and two other customer-specified conditions. Any reading can be shown or all readings can be scanned continuously. In addition to the front panel display, monitoring of the functions is also done at a remote-monitoring panel in a nearby office, using a 2-wire twisted-pair communications circuit.

For long-term economic benefits, the PDU offered the best solution. The presently adequate 30kVA unit can be paralleled in the future with one or more additional modules to gain increased capacity. The unit is comparatively quiet (55 dBA average sound level measured at 5 ft), and at three-quarter load would produce 5800 Btu/hr, which could be easily handled by their existing computer room air-conditioning equipment.

The firm was also concerned about loss of utility power. Only enough energy is stored in a magnetic synthesizer to ride through about a one cycle outage (16.7 milliseconds). A full UPS, however, was not initially deemed necessary since the law office is in an area that has a highly reliable utility network — only a single power outage in more than 20 years. Therefore, the PDU was expected to solve 99.5.% of any computer power problems.

DISTURBANCE ANALYZERS

Some facilities containing solid-state electronic equipment have complex distribution systems that are highly susceptible to power surges and sags. Rather than conducting tests by using portable instruments, these facilities opt for permanently connecting a series of disturbance analyzers at critical locations in the system in order to track the stability of their system.

Case history 1. Because of the number of points requiring monitoring, at one hospital the maintenance department has installed an automatic monitoring system that features electrical disturbance analyzers at remote monitoring points and standard telephone cable linking them to a central computer.

Prior to the installation of the new system, electrical loads had been monitored by a single portable analyzer. Due to the quantity of requests for information on the various distribution systems, the single analyzer could not handle the demand. A detailed analysis of the hospital's requirements indicated that a new system was needed that was capable of monitoring multiple electrical systems simultaneously while providing real-time performance reports and maintaining a historical data base for trend analysis.

As shown in **Fig. 12.7**, the multipoint monitoring system selected contains sev-

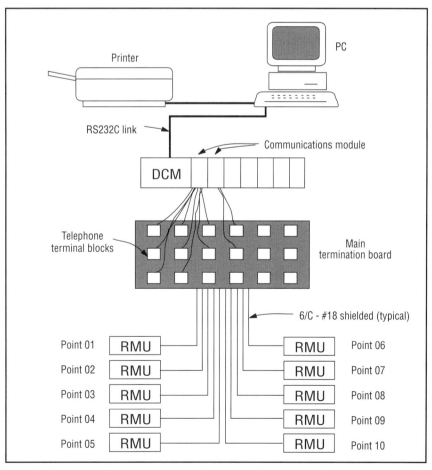

Fig. 12.7. *Automatic remote monitoring system showing the data communications multiplexer (DCM) and remote monitoring units (RMUs). The system required no dedicated computer, so an existing PC was used.*

eral remote monitoring units (RMUs) and a data communications multiplexer (DCM) linked together by telephone cable.

Each RMU is a power-line disturbance analyzer capable of accepting programmed thresholds, and measuring and recording power line sags, surges, high and low voltages, impulses, and neutral-to-ground voltage peaks. Stored event information includes date and time of occurrence, the phase in which the event occurs, type of disturbance, amplitude, and for sags and surges, the duration.

Linking the various RMUs to the computer is the DCM. Within the DCM are separate plug-in type communications modules, each having eight channels with RJ11 ports. Depending on the signal type and carrier speed required, different modules can be selected. In this instance, 300 baud FSK-type signal modules are used. The system can be expanded to 64 remote-monitoring points by adding additional modules.

A fused terminal block is mounted at each monitoring point. Individual neutrals as well as phase conductors are run from the monitored panel, through the block, to the RMU. The neutrals are included since neutral-to-ground voltage levels are monitored and spikes are alarmed.

Information is then transmitted from each RMU via shielded telephone cable to a main termination board. Shielded cable is used to avoid interference from 277V and 480V AC systems, especially when these cables occupy the same plenum space. Mounted on the board are telephone terminal blocks, one for each RMU. Standard flat telephone cable connects each terminal block to the DCM, which is located in the electric shop office. The farthest distance from an RMU to the DCM is about 2600 ft.

Linking the DCM to the computer, also located in the electric shop office, is a RS232C-type connection. An auxiliary port is provided for future connection to another group mainframe.

A variety of bar graphs coupled with an alarm status/main menu display are provided for analysis and alarm annunciation of the power distribution system. The following items are tracked.

• The number of disturbances of a selected type occurring during a defined time period correlates local problems with generalized power disturbances. A printout of the events shows date, time, monitoring point number, channel (phase or neutral), event type, voltage, and duration.

• The number of disturbances of a selected type at a selected point, within the time interval they occurred can be displayed. A printout of the events generating this display shows detailed information per event.

• The quantity of disturbances within a 24-hr period of a specific group of monitoring points can be displayed. A printout of the specific events detail each occurrence.

An alarm status/main menu provides an instantaneous overview of conditions at critical monitored points. Each annunciated point display shows RMU number, point, number, and location. Below the status display is the main menu from which specific system monitoring parameters can be selected. Upon selection of the monitoring point and parameter, the desired information appears on screen.

The original portable disturbance analyzers are currently used to monitor specific loads, such as radiology. For example, if an X-ray machine is suspected of putting interference back on the line, a portable analyzer is connected between the main power source and the X-ray equipment, and a check for spikes is conducted. The analyzer will then indicate from which direction the spikes originate thereby determining whether the problem is power source or equipment related.

Case history 2. Transformer overheating and excessive noise, burned-out connectors in prewired office partitions, and overheated neutral conductors occurred in an electrical distribution system that, on the surface, seemed to be operating well within the ratings of all equipment. The problems occurred on the parts of the system feeding concentrations of office PCs and other microprocessor-based devices. Preliminary checks made with true-rms reading ammeters indicated that the system were reasonably well balanced and that the loads were clearly within the capacity of the transformers, yet overheating still occurred.

The problem was particularly evident where the offices and work areas were comprised of preconstructed flexible modules. Here, the modular wall partitions are prewired; and as they are assembled, the wiring is connected from one partition to another using plug-and-socket connectors. In several instances, these connectors burned out from what appeared to be steady-state overload conditions. The connector discoloration and lack of evidence of arcing indicated long-term heating, rather than rapid short-circuit failure.

The installation uses a 480V delta to 208Y/120V step-down transformer to supply the 120V phase-to-neutral loads. The neutral is common to all loads and is the same size as the phase conductors. The transformer is rated 112½kVA. As shown in the table of **Fig. 12.8**, the loads are reasonably but not perfectly balanced among the three phases, and the transformer is loaded well below its rated full-load current.

The load is mixed, including electronic office equipment and conventional loads. PCs and peripheral equipment almost all use switch-mode power supplies that draw power from the source in pulses, rather than continuously, resulting in severe nonlinearity and high harmonic content.

Since the load included numerous switching power supplies, it was suspected that the apparent overload conditions and noise problems were caused by distorted current waveforms with high harmonic content. Tests were run using a disturbance waveform analyzer

	Actual current	% total load
Phase A	257A	31
Phase B	272A	33
Phase C	298A	36
Average 3Ø	276A	

Fig. 12.8. *Load balance: full-load current for 112.5 kVA transformer is 312A, and measured value is 88% of full load.*

to obtain current and voltage waveform traces, with a PC and software used to print out the harmonic analysis in graphic and tabular form. **Fig. 12.9** shows the transformer secondary current and voltage waveforms obtained. These are typical for a mixture of linear and nonlinear loads, as in most office environments.

Fig. 12.10 shows the harmonic current in each phase and neutral as actually measured by the harmonic analyzer. For the phases, the harmonic distortion is given in percent of the fundamental current, which is customary, since each harmonic is normally less than the fundamental current.

For the neutral, however, the harmonic current is given in rms amperes, for two reasons.

The phase currents are reasonably balanced, so the 60-Hz fundamental current in the neutral is small, but the harmonic currents are added (vectorially) and are relatively very large. If expressed in percent, the total harmonic current, mostly third harmonic, would be about 750% of the 60-Hz current.

The neutral current in amperes gives an indication of the amount of neutral heating that will occur. Note the high third harmonic content and the smaller contribution of the ninth harmonic, both of which are arithmetically additive in the neutral. The total harmonic current of 228A in the neutral is almost exactly equal to the rms amperes shown on the neutral current waveform in Fig. 12.9(C), indicating that there is almost no fundamental frequency current carried by the neutral. This is as expected for a balanced load. The neutral current is almost totally 180 Hz (third harmonic).

The crest factor of a pure sinusoidal waveform is 1.414. For the phase currents, neutral current, and Phase-A voltage in this case, the crest factors are given in **Fig. 12.11**. The high current crest factors are typical of waveforms distorted by current drawn in pulses by switch-mode power supplies. The voltage waveforms, obtained directly at the transformer secondary, has a crest factor in the range of 1.35, which is less than that of a pure sine wave. This is because the high pulse current causes saturated transformer conditions, and the voltage wave is flat on top, never reaching its true sine-wave peak. Because the load is mixed, the flattening of the voltage wave is not very pronounced.

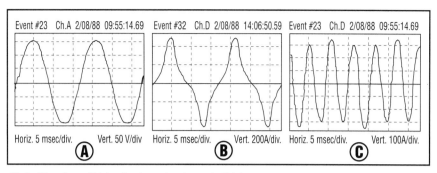

Fig. 12.9. Waveform (A) is of voltage in phase A; (B) is current in phase A (current in phases B and C are similar except displaced by 120° and 240°); and (C) is current in neutral. Phase current is basically 60 Hz, but highly distorted; the neutral current is predominantly 180 Hz (3rd harmonic).

Harmonic	Phase A%	Phase B%	Phase C%	Neutral (Amps)
3rd	30.26	33.66	33.39	227.75
5th	8.55	10.74	10.71	3.63
7th	1.92	2.88	2.94	3.90
9th	1.46	2.03	1.94	12.06
THD	**31.53%**	**35.51%**	**35.24%**	**228.17A**

Fig. 12.10. Harmonic current measured. Total harmonic distrotion (THD) is vector (not arithmetic) sum, allowing for phase angle.

Fixing the burned-out pressure connectors in the prewired modular partitions was straightforward. The plug-in connectors were removed and the partitions hardwired. This eliminated the problem of plug-in connector overheating.

One solution for the branch circuit wiring used in the partitions, is to use a separate full-size neutral for each phase conductor. This provides three 2-wire (phase and neutral) branch circuits, rather than a 3-phase, 4-wire multiconductor branch circuit with a single common neutral conductor. In the 2-wire branch circuit, each neutral carries the same current as the phase conductor, not the vector sum of all three phase conductors carried by the common neutral. If the rms current (including harmonics) in the phase conductor does not exceed the conductor rating, neither will it exceed the rating of the full-size neutral.

Another possibility is since the total rms current, including harmonics, in the common neutral of a 4-wire circuit can be a maximum of 1.73 times the phase current, for simplicity making the neutral double the ampacity of the phase conductors. This could be done for the 4-wire branch circuit instead of using three 2-wire branch circuits.

The second major problem was the overheating of transformers. Standard transformers are designed to handle 60-Hz power, and they do not take kindly to higher-frequency harmonics. Currents of 180 Hz and higher circulating in a transformer cause greater magnetic hysteresis and eddy-current losses in the iron of the core and a more pronounced

	Peak/RMS	Crest factor
Phase A current	475/257	1.84
Phase B current	510/272	1.88
Phase C current	557/298	1.87
Neutral current	350/229	1.53
Phase A voltage	154/113	1.36
Average current (3 phase)	**514/276**	**1.86**

Fig. 12.11. Crest factors.

skin effect in the winding conductors. Also, they may cause saturation of the transformer steel, resulting in large increases in current with small increases in voltage on the transformer. These factors result in higher operating temperatures, insulation degradation, and a substantially shortened life. Sometimes the excess heating is sufficiently severe to cause transformer failure. Also, the harmonics cause a considerable increase in transformer noise, especially if the core becomes saturated.

Rather than replacing the existing transformers with k-Factor ones, the transformers were derated to prevent excessive heating and to reduce noise. A commonly used rule-of-thumb was used to calculate a derating factor. It is obtained by dividing the peak-load current if the rms current were a true sine wave (1.414 times the rms current) by the actual peak current.

Per Fig. 12.11, the average 3-phase current on the transformer is 276A (about 88% of full load). Actual average-measured peak current is 514A. Using the rule-of-thumb, the transformer was derated to a load factor (percent of full load) as follows:

276A × 1.414 = 390A true sine-wave peak.

Actual average measured peak = 514A.

Load factor = 390 ÷ 514
= 0.76 (76%).

The actual effect on transformer heating will depend on the characteristics of the individual transformer and the specific load.

LOCATION CASE HISTORIES

This series of actual case histories serve to illustrate the varied problems found in different facilities. They illustrate that when called in to evaluate the lack of quality power for the solid-state electronics installed there, one must not go to the jobsite with preconceived notions of what will be found or solutions that can be applied.

OVERNIGHT SHIPPER

One of the modern miracles in a small package is called the network, which is basically a series of powerful workstations tied together with a central unit called a server. All of this was made possible by the high-speed, high-capacity, intelligent personal computer (PC).

Fig. 13.1 is a simplified diagram of a local area network (LAN) used in a billing operation by a major overnight air shipper. This system serves a total of 17 operators at workstations processing the way-bills from outlying offices in one region of the shipper's nationwide operation. The processing is vital to the issuing of 20,000 to 40,000 invoices to customers throughout the region and obviously represents a sizable amount of income.

At the end of the radial lines are operator workstations, all supported by the server at the hub of the system. The server's operation is the most vital link in the network because it is common to all operators and is the link to master logging of information at company headquarters. In this case, the link is over a larger LAN joining together the regional centers across the country. Because of its importance, the server is normally powered by a UPS.

All of this equipment is single-phase, 120V, and connected with 2-wire (plus ground) plugs and receptacles. The ease of connection tends to lead installers to assume they do not need to be very concerned about protective practices. Such things as ground integrity for power and signal referencing, transient surge protection, and powering of a harmonic-producing power supply tend to be overlooked since the main server

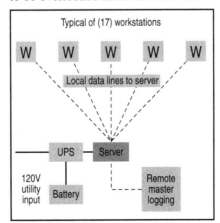

Fig. 13.1. Local area network wiring.

is connected to a UPS, and the workstations into the wall receptacles.

This client, an air shipper, had a national contract with a software and hardware support company to troubleshoot and correct operating problems directly with each site. In addition, the air shipper had contracts with local electrical contractors to perform maintenance and wiring changes whenever needed. The hardware-software support company was the supplier and installer of the computer equipment, including the UPS supporting the server.

We were called to examine this site with the complaint that the billing operation was intermittent and invoices were not getting out in timely fashion. When we arrived, the local electrician was able to show us the power company service, panelboards, grounding, and service distribution. We also were told the power company was changing out some of their transformer and distribution equipment outside the building due to this client's complaint that the power was of poor quality! Interestingly, there were three other computers running with no conditioners or UPSs on the supposed poor power.

When we came to the network room, the supervisor voiced concerns about the continual warnings showing up on the UPS display that the power to the network was "about to go away." This display lets the operator know how much time is left on the battery system so the operation can be brought down without loss of data. In this case, this warning was coming repeatedly, and the operators were following the warning by closing down their systems each time. The supervisor advised us that the red light was almost always ON. Reading the UPS manual, we found that the red light was an indicator of battery charge or battery loss (when the light was on continuously).

At the end of the day's processing, so as not to create any further delays, we offered to test the whole UPS protection system by pulling the input plug to simulate actual utility loss. We suggested what the supervisor was seeing was a bad battery that was no longer able to supply energy when the utility power failed. When we pulled the plug, the entire system crashed, proving the expectation.

What can we learn? A little understanding of the basic operation can lead to practical housekeeping. In this case, the UPS unit was installed to maintain power during voltage sags, surges, and outages. The problem was not the UPS but the apparent gap in the responsibility chain: one man installed what he did not understand and the electrical maintainer had no responsibility to check on the action of the product, since he did not provide it.

The solution: Be knowledgeable of operation instructions and let all the participants have a say-so in developing the rules.

FROZEN FOOD PROCESSOR

In certain instances, system operation problems can be intermittent, coming and going without pattern or apparent correlation. This usually indicates a wiring and/or grounding problem allowing multiple pathways for circulating electrical noise to exist. As a result, the noise can interfere with a sensitive signal circuit. Because the signal circuit is not protected from such phenomena, it might become disrupted or damaged by this transient energy.

A medium sized industrial customer in the food processing industry called because it was experiencing operating problems with conveyor systems that moved product through the preparation process. This equipment was powered by a group of three 5 hp variable-speed drives that had recently been installed. Drive system speed depended upon the level of production. Disturbances to the conveyor systems were so severe that the system controls had to be manually reset in order to continue production.

The facility manager, who was familiar with the problems associated with harmonics, believed there was a harmonic interaction problem on the electrical distribution system, primarily due to the new drives. Further discussion revealed that the building service transformer was rated at 3000kVA and that the three drive systems were the only harmonic producing loads in the processing area.

Doubting seriously that harmonics interaction was the cause of the trouble (because of the small number of non-linear loads relative to the electrical service size), we suggested that a further survey of the site might provide some additional information.

Voltage and current measurements were taken with a true rms digital multimeter at the building service and found to be well within specified operating parameters and capacity. Using a harmonic analyzer, we observed that the THD (total harmonic distortion) in current was less than 2% and individual harmonic distortion, in current, for the odd harmonics up to the 21st to be less than 3% each.

A study of the utility distribution system operation over the prior three months did not reveal any correlation between its operation and the upsets experienced at this site.

We then moved to the drive locations to examine the installation. Each drive system had its own dedicated circuit from a subpanel and each of these circuits fed a 120V/120V, single-phase, electrostatically shielded, wall-mounted transformer near each drive. The transformer's secondary then fed the drive control system.

From a visual standpoint, it was apparent that the installer had given thought to the installation by using dedicated circuits and shielded transformers. They assist in the control of electrical noise while establishing a solid ground reference for the electronics package referenced to ground.

To verify that the transformers were indeed providing noise rejection and a stable ground reference, we again used the true rms multimeter to measure voltage between the case of one drive control system and the power system neutral. You would expect to measure zero volts at this point. Instead we measured a power system neutral-to-case ground voltage of 35V! When the other two transformers were investigated, approximately the same voltages were measured. At least we were on the track of something consistent with the possible explanation of the system's behavior.

Fig. 13.2 shows the proper installation of these transformers. Notice how the incoming ground, shield, core, secondary neutral and secondary ground are all bonded at one point. This is called a single point ground. This single point ground is then grounded to the building ground system.

When the covers of the transformers were removed to verify the wiring

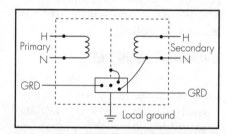

Fig. 13.2. Proper wiring of shielded isolation transformer has incoming ground, shield, core, secondary neutral and secondary ground all bonded at one point (a single-point ground), which is then grounded to the building ground system.

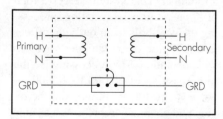

Fig. 13.3. Improper wiring of shielded isolation transformer has neutral not bonded to the single-point ground; also, transformer is not locally grounded to the building ground system.

connections, what was found is shown in **Fig. 13.3**. The neutral was not bonded to the single point ground and the transformer was not locally grounded to the building ground system. When these connections were made, the neutral-to-case ground voltage returned to zero and the unexplained upsets ceased to occur.

This provides an interesting example of the difference between a power-related anomaly, or something that repeats with a cyclical pattern, and a disturbance that behaves in a random or transitory fashion. Here, the end user had not given any thought to wiring-related problems, assuming instead that the equipment installer had taken care of the necessary details. Mistakenly, the transitory disturbances were attributed to harmonics.

TELECOMMUNICATIONS MANUFACTURER

Most solutions to power quality related events come from attending to little things. In almost every case involving a power quality investigation or site analysis, the first items in need of correction are wiring and grounding. These include daisy-chaining of neutrals and grounds, undersizing of ground wires (which are expected to carry electrical noise away from sensitive equipment), and even loose or missing connections.

In the context of power quality and the NEC, there are conflicting philosophies promoted by people with differing technical backgrounds and commercial interests. This apparent conflict can only be settled by making safety the overriding concern, which is the intent of the NEC. If revisions have to be made for the system to operate properly, the equipment manufacturer must incorporate them in the equipment design, rather than asking for deviations from the NEC. In fact, when wiring is installed with NEC safety provisions in place, satisfactory operation will not be sacrificed. The safest system also will be the best operating one.

Good housekeeping is not just following the NEC, but includes all the sensible practices taken during installation and maintenance. If more attention is given at the start of a project to the important power quality implications, future problems will be headed off before they result in shutdowns or damage.

Many times, we're involved with a consultant, contractor, or facility engineer who has not thought through the hows or what-ifs pertaining to wiring, grounding, or protecting a piece of solid-state equipment or its signal system. Thus, we are called in when this equipment doesn't work, or when there is no correlation to the errant behavior, or when the damage has already taken place. One case comes to mind as a reminder to check procedures for installing and maintaining equipment, especially power conditioners used on computers or other solid-state electronic loads.

A large telecommunications facility that manufactured digital electronic devices was experiencing a rapid increase in requests for UPS systems from operators in its software engineering area. This was the direct result of one software developer asking to have a small electronic equipment system (4 to 6kVA) put on a self-contained battery supported UPS. When this request was honored and a 10kVA UPS provided, the developer had clean power since his work area was now separated from the wiring problems in the rest of the software engineering area.

As word spread among the other personnel in this department, more requests were made for small UPS systems, so many requests, in fact, that the facility manager had to put a stop to the separate ordering until it was determined why everyone needed a UPS! The answer from the software engineers was, "You're giving us bad power!"

We were called in to do a site analysis at this facility and, in the course of the visit, we tested the output of one of these small UPS systems to determine its load characteristics. We were performing our examination with the load connected and in a nonintrusive fashion, or so we hoped.

We were suddenly confronted by a developer wanting to know what we had done to her computer! It was now "dead in the water," and she was suppose to go home in 30 min. Thanks to us, however, she now had a 6-hour recovery program to go through to retrieve the lost data! Needless to say, our credibility was suspect.

We re-checked the output of the UPS. All indicating lights showed good output, the voltage was correct, but no current was present. Then we went to the respective power outlet and found the same condition. The obvious theory: something was wrong between the UPS, which was running well, and the power outlet. The operator maintained, "It's you; nothing else has happened except when you tampered with the system!"

We returned to the UPS output to prove that we had not changed anything and that the power was correct at the output. When we placed the voltmeter leads on the output to show her the voltage, one of the output wires moved sideways.

Looking underneath the connection into the lug position, we saw that the outgoing wire had never been inserted into the lug. In fact, the lug had never been opened! The wire had simply been jammed up to make contact, and as long as the arrangement was not touched, it worked. The rest of the units were checked for the same problem, and better planning for installation and maintenance was instituted for the future.

OFFICE BUILDING

An 18-story office building under construction was designed with two power risers for distribution to the various floors, with each riser having its own step-down transformer. One riser fed the first and second floors as well as the top floor, which contained five

Building management systems *are increasingly being used to monitor and control power distribution, energy management, and HVAC systems from a central location.*

131

elevator controllers fed from an 800A power distribution panel. The other riser fed the third through the 18th floors.

As construction progressed, energy control devices were installed at each floor and tested. During testing, the devices installed on the third through the 18th floors worked fine while those installed on the first and second floors were damaged when tested. In other words, the units powered by the riser serving the elevator controls were negatively affected.

The energy controls supplier thus went to the building owner, complaining that its contract did not include replacing units damaged by "bad power from the elevators." The owner responded that no such thing was happening and that the "defective" devices had to be replaced at no cost. We were called in to verify what was happening.

Suspecting something was happening on the first power riser, we proceeded to the top floor to examine the currents flowing in the elevator service panel and the individual control devices for each elevator.

SCR control equipment was being used to supply smooth, adjustable power to the elevators, with an eye toward energy conservation. Several "snapshots" of the current spectrum were taken to determine the extent of distortion in current flowing from these 6-pulse controllers.

The total current from the elevator service panel back onto the riser bus was extremely distorted, showing a large content of high frequencies in the total rms current. Specifically, 60% of the fundamental was in the fifth harmonic, 40% in the seventh, 15% in the 11th, and 5% in the 13th. When the individual controller currents were examined, similar percentage readings were observed, with very little canceling effect from one device to another when several elevators were running.

The problem obviously centered on the interaction between the elevator controllers and the power riser step-down transformer's impedance. A large amount of distorted current flowed into and interacted with the impedance of the step-down power transformer serving the power riser. The result was a significant voltage distortion on the power riser that, in turn, caused damage to energy controllers on floors one and two.

It's interesting to note that the interaction did not affect the second power riser because it was reduced by passing through both step-down transformers (two inductances), and thus did not effect those energy controllers on floors three through 18.

Apparently, little effort was made in the design to contain or mitigate these harmonic currents coming out of the rectifier inputs to the drive systems. Necessary corrections had to be made at the source: the elevator controllers.

The elevator manufacturer was contacted for a recommendation on methods to limit the amount of distortion coming from its units. The time was past for changes such as installation of 12-pulse units that would eliminate the 5th and 7th harmonic currents. The chief engineer of the elevator company recognized the problem and agreed to review the impedance of the circuit so that properly tuned filters could be added on the elevator controllers.

In this instance, much attention in the new drive system portion of the equipment spec was devoted to the operating end of the system, the place where the fine control, smooth running, and excellent adjustability take place: the drive motor. While this attention is well deserved (since this is the purpose of the controller), it's equally important to carefully plan the connection to the power distribution system. In many systems, the conversion process at the input is given so little attention that considerable distortion demands are made on the power connection.

DIE MAKER

Disturbances traveling through the air are not as widely known and understood as those traveling through power and/or voice/data wiring systems. In addressing power quality problems, we usually concentrate on what we might call the major areas of concern: grounding, wiring, transients, harmonics, and power conditioning. Nevertheless, disturbances traveling through the air represent the same potential for trouble and we must be aware of their presence.

We were called to a manufacturing facility that produced master patterns for die making. The request came based on the owner's initial assumption that bad power was energizing an expensive new cutting machine, which used submerged-arc laser cutting technology in order to attain tolerances down to 1 micron. The owner had complained to the machine supplier that exhibited tolerances were not in accord with specifications, that precision was poor, and that the machine needed to be fixed or it would be returned.

Since this installation was so expensive and the equipment manufacturer did not wish to consider having to accept a return, a team of service personnel came to the site to make whatever corrections were necessary. After they had tested the piece of equipment to verify its operation, they concluded that the problem was not with their machine but rather that energy was "leaking" from the power company transformer, which is located on the outside of the wall of the room where the cutting machine was running. Thus, the owner called us in to perform a site analysis to verify the equipment manufacturer's contention.

The location of the sensitive cutting machine, as shown in **Fig. 13.4**, was in a specially prepared room, where site preparation procedures had been practiced to protect the machine from any disturbing influences. Based on the

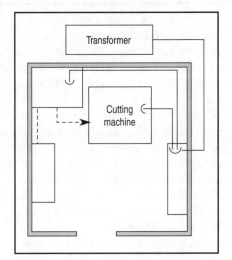

Fig. 13.4. Supplier alleged that energy was leaking from the utility transformer and affecting the operation of the cutting machine.

equipment manufacturer's statement, we brought with us a field strength meter, as well as a power disturbance analyzer and other instrumentation.

Grounding and wiring connections were examined and found to be made well; they would not support ground loops that might upset the cutting machine's sensitivity. We then evaluated the machine's performance relative to any transient events and found that the problems had very little to do with these events. Rather, the performance problems happened somewhat regularly even without storm or switching surges. Neither could the problem be tied to any fluctuation or intermittency of the power supply and did not seem to have any correlation with harmonic currents being created by the machine power supply.

We repeated several of our meter hook-ups, with the machine idle as well as running, and could not find any difference in the readings.

Field strength meter measurements were taken in the room where the machine was operating; no field strength voltages of any consequence were found. The same was done outside, in the vicinity of the power company transformer. Interestingly, we found that what little field there was to measure was in the opposite direction from the building. Even though we had satisfied the owner that the power company was not leaking power into the equipment, as was contended, we had found no solution to the problem.

To be sure we had not overlooked some simple point, we asked the owner to walk us through the entire process, from start to finish, of how an order to make dies for a customer was handled. Since we had already seen the actual cutting operation and examined the physical layout of that room, we went to the other areas just to be sure of all our surroundings. We looked at the incoming order department, the quotations department, the materials handling area, and finally the engineering department.

At this point we asked a simple question. "How does the design data for the operation of the cutting tool to make a specific die get to the cutting machine? We presumed the design area comput-

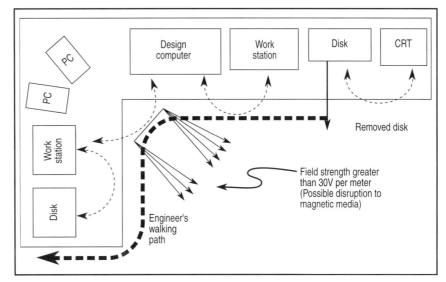

Fig. 13.5. *Electromagnetic field from workstation equipment damaged the disk hand carried by the engineer.*

ers had a communication link to output the logic commands directly to the tool for operation.

Yes, in fact, there was a link, but not what we expected! The configuration of the computers in the engineering department where this software was prepared is shown in **Fig. 13.5**. At an L-shaped group of PCs and workstations where the assembly of the software was made, the final design criteria were stored by an engineer on a 3½-in. disk. From here, the disk was hand-carried out of the area to the cutting machine.

We decided to duplicate this link, this time carrying a field strength meter. We followed the engineer with our meter as he left his workstation area. As we passed the corner of the L-shaped work tables, carrying our instrument sensor at the same level where his hand had been, the meter suddenly jumped to read over 30V per meter of field strength! While we were stunned by the discovery, we all agreed we were probably looking at the answer: a disturbance through the air to the finished cutting design data stored on the disk. The owner agreed to change the link pathway and to protect the logic contained on the disk as it progressed to the machine. In subsequent cutting machine tests, the sensitivity was restored to the operation.

TEST FACILITY

Are there ways in which electrical noise can be injected into wired circuits or sensitive equipment? If an answer of "no" is given, we would have forgotten about the electrical and magnetic fields associated with 60-Hz currents, and how they might couple their influence into other metallic circuits. Also, the effects of different frequencies, especially the high frequencies associated with high-speed transmissions, must be considered.

Many site visits and case studies have shown that forgetting the different means of injecting noise leads to weak links in our noise rejecting and protecting methods.

Several years ago, we were asked to help to make a final test facility keep running. We examined the conditions that were needed to have an uninterrupted four-hour period of operation in order to validate the final production test for the product. The facility was bothered by power quality problems, which were assumed due to unstable power delivery from the local electric utility. On this assumption, all kinds of power conditioners had been bought including voltage stabilizers, line conditioners, even a small battery-supported UPS.

If the problem was on the power line, that approach would have been correct. But it turned out the disturbances were coming out of the ground reference for the digital signal drivers and receivers in the test circuit. Thus, all of the cures (in the form of power conditioning equipment), while working for line-to-line noise, did nothing for common-mode noise.

We disconnected the client's test equipment and made temporary coils in the cord sets to see if we could stop any high-speed traveling spikes on the grounding conductor. As we suspected, the disruption stopped, and the test equipment did its job while running without interruption, even with all the power conditioning devices removed.

Notice that this is a neat diagnostic trick, even if you don't see the same results seen in our example here. What do we mean? Well, if you do see the same kind of improvement, you know that the disturbance is traveling down the ground wire. You then use a transformer or a balun as a permanent fix.

But would you learn anything if you did not get results? The answer is YES; you would have settled the issue of noise spikes coming down the wires. In other words, this method gives you an answer in either case. Either you know what's there and can fix it, or you know what's not there. If no results are obtained, you don't need to continue to look for the wired-form of noise, but can look for some other means for the noise to be disturbing the host equipment. This is invaluable as a quick check on where to start.

COMPUTER COMPANY

Ground integrity and noise rejection are like the two sides of a coin and almost as inseparable. When we stop to reflect on the difference between safety grounding and signal protection, we can see an enormous gap between safety and top notch operating performance.

With ground integrity, we are trying to establish a stable ground plane of equal voltage (equipotential plane) for the signal circuit. If we don't have it, the signal will rise and fall with whatever fluctuation occurs that affects the ground reference plane. When the plane of reference is not disturbed by changing external conditions, then the signal is stable and readable by the sensitive drivers and receivers of a signal system.

Remember that the chassis or case of electronic equipment, which is connected to electrical safety ground, is what is used to "reference" the signal drivers and receivers. As such, we are concerned with how, not if, this connection is made.

With noise rejection, we are employing the means to keep incidental unwanted signals from interfering with the desired signal. But this is done once we have the desired signal referenced to a stable ground plane.

We were called into a facility to determine why there was "no ground." The user was sure that a ground wire was pulled in conduits and connected to receptacles. In fact, it had insisted on this fact for over a year when defending its installation to a computer company's site preparation engineer. This engineer, in turn, was at a loss to explain why tests of the computer company's system indicated no ground reference.

We proceeded to use a ground circuit checking tester and found that the tester indeed indicated "no ground" as the computer company engineer had claimed. Investigating further, we removed a receptacle faceplate, pulled the receptacle out, and were amazed to find that the green wire was there all right, but coiled in the bottom of the receptacle box unconnected. The ground wire connection was then made here and at all other receptacles involved. The problem remained. The obvious question became, "What about the panel end?"

We went back to the panel serving this group of 20 branch circuits to see if we could figure out why the engineer still could not get a decent ground reading. Remember, it's the computer engineer's responsibility as an installation supervisor for his company to certify the signal reference stability for the company's computer. When we reached the panelboard and removed the panel front, we found all the ground wires dangling and unconnected.

Remember, there are two goals for grounding: the challenge is to make sure of both. It must have:
- fault return paths to trip circuit breakers upon fault conditions; and
- noise-protected, signal reference connections to eliminate upsetting ground loops.

AUTOMATED WAREHOUSE

All electrical installations require a good grounding system for personnel and equipment protection. Computer systems are particularly sensitive to transient signals that can be fed back through their digital circuitry, and errors can result. Because of the delicate physical makeup of circuit boards and the logic elements they contain, computers are also subject to internal damage and component failure. A good grounding system minimizes these hazards.

Many installations go wrong in grounding their PCs and other electronic equipment. It is essential that neutrals and grounds be segregated within IG receptacles. There should be only one path of lowest impedance to ground via the most direct path to the transformer array ground. Any alternative paths, including routes through the communications ports, circuit boards, and computer power supplies, can cause the mysterious burnout failures that are often blamed on poor equipment.

We were called to an installation consisting of a highly automated warehouse and an office building housing a computer center, which had 135 online order terminals. Three distinct computer systems were involved. One handles the order entry, label generation for the warehouse order fulfillment system, and billing. A second computer and peripherals were dedicated to gathering historical data, providing customized pricing, strategy information, and inventory monitoring. These two systems were located in the office building. The third system, located in the warehouse, was used for controlling the operation of conveyor systems and other automatic equipment that transport items to trucks waiting at the loading docks.

This site unintentionally provided a comparison between good (the computer center) and bad (the warehouse) computer power installation techniques. After commissioning the warehouse system, many repetitive failures occurred on the approximately 135 video display terminals scattered throughout the warehouse area. Additionally, order-taking and shipment programs, which were interrupted almost daily because of power-related problems, had to be rerun, resulting in delayed shipments. Our assignment was to track down the source of the problem and recommend corrective action.

Many cables connected the computers in the office complex and warehouse terminals. One of the first things noted was that no consistent grounding scheme had been followed by the installer in the warehouse. While IG receptacles had been installed, ground wires in the receptacle boxes were either nonexistent, bonded to the box, or tied to the neutral terminal of the IG receptacle. In some cases, the ground wire was run in conduit up to and connected to the roof-truss steel. The multiplicity of ground-return paths encouraged the circulation of ground-loop currents. Lack of a single-point ground reference allowed surges induced by lightning to seek the lowest-impedance path to ground, which often was through computer peripheral equipment.

The only columns in the warehouse were internal ones that support the roof trusses. Because of the huge size of the warehouse, data cables running from the computers in the office building to the terminals in the warehouse were strung along the roof-truss steel. Without a solid grounding system, the roof steel acted as a giant lightning rod and antenna. Sneak paths for induced currents within the cable shielding were causing much of the erratic computer operation. Adequate grounding of roof steel thus becomes important.

Initial recommendations for improving the situation primarily focused on the grounding system and installing surge protective devices similar to those used in the computer center. Because this is an operating facility, correcting the receptacle ground-wire situation proceeded on an "as available" basis.Similarly, cable strapped to roof steel was rerouted as time permitted. To answer the immediate need, small individual surge-suppression protective devices were installed at all major pieces of equipment.

Dramatic improvements in the reliability and life of the warehouse equipment resulted from the relatively simple approach taken. The failure rate of the equipment is now virtually identical with that of equipment located in the computer center.

The major lesson learned from this is that more is not necessarily better. While it may seem that there can never be too much grounding, this may not be true where electronic equipment is involved. Better grounding rather than more grounding is what is needed. Also, computer power, grounding, and surge protection should be treated as an entity.

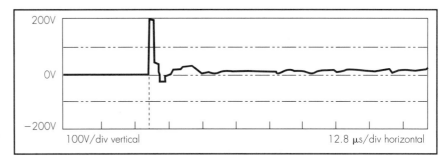

Fig. 13.6. Trace of high-frequency surge (transient) occurring close to measuring device. Very steep wavefront rise reflects close proximity.

HOSPITAL

There still are many who assume their problems are caused by their power source. Usually, these are people who have no power conditioning equipment but think they need it to protect their sensitive devices from a disturbance event. And, it's very difficult to convince them that concerns other than power may be equally important. Our standard question to a client asking for an estimate on a UPS system is, "If you install this power device with battery backup and it does not solve your problem, what will you do next?" The standard response is astonishment: "How could this much expense not solve the problem?"

To verify if the disturbance is power-related, first determine whether it correlates with any power-related event at the facility or on the power lines serving the location. If the correlation is supported by a time stamp of power-related events, you can investigate the nature of the event and consider available alternatives. These may include separating the electronic equipment from the disturbances or mitigating the disturbances with a power conditioner.

If, as is more often the case, there is no power-related event observed at the time of the upset, consider other conditions as prime candidates.

A hospital wanted answers to why its new variable air volume (VAV) controllers were misbehaving when mysterious spikes hit its electrical system. We connected a monitoring device to the power distribution system and obtained the trace shown in **Fig. 13.6**. This appeared to be a very high-speed surge peaking at well over 200V when measured on a system that could stand only 20V to 30V peak before the circuit was damaged.

The client looked at this spike and immediately concluded the electric utility had transmitted this disturbance on its power line. We explained that while there may be certain conditions that could be blamed on the utility, this certainly was not one of them. We knew there was no known power activity taking place while we were making the measurement, and we sought to verify this by contacting the utility. Sure enough, the utility confirmed that no storm activity, sagging, or circuit breaker operations took place at the time in question.

Look at the trace closely to see if more information can be derived. First note the time scale: Not seconds, not milliseconds, but microseconds! We are looking at 128 microseconds, or one-eighth of one millisecond, for the entire trace as shown. Second, note that the wavefront of the spike rises almost straight up vertically. With the above analysis of the trace, it's obvious that it would be physically impossible for the energy represented by this spike to have traveled over any distance. If the

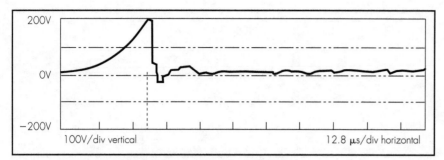

Fig. 13.7. Trace of high-frequency surge with sloped line indicating what trace would be like if surge traveled from a distance.

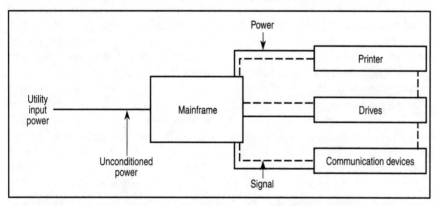

Fig. 13.8. Mainframe computer is connected directly to the utility input power line.

spike travels a distance, it would look more like the one shown in **Fig. 13.7**. Time (represented by the sloped line) would be needed for the transmission of this disturbance. This was the confirmation for rejecting the theory that the utility was the source. The problem source was obviously nearby.

Upon investigation, we found arcing in a motor connection on one of the hospital's elevators. To protect the VAVs, we not only recommended repair of the elevator motor connection but also suggested the use of in-line reactors or chokes ahead of the VAVs to soften any high-speed surges.

Note that what started out as a power supply concern turned out to be a solution targeted inside the facility.

MECHANICAL CONTRACTOR

A mechanical contractor installed microprocessor-controlled compressors in his workshop, some with variable speed drives and some with across-the-line starters. Convinced that the power company was delivering bad power, the contractor was ready to order power conditioning equipment to smooth out any disturbances affecting the microprocessors.

To examine the site for the type of equipment upsets, we connected a disturbance analyzer on the incoming power lines. We noted only four or five incidents of increased voltage in a 2-week period, and then only 2% to 3% above our threshold. This certainly did not warrant a power conditioning device. And, we found that the compressor equipment and microprocessor controllers were erratic whether the drives were on or off.

With further investigation, we discovered one problem: The lack of a stable ground reference for the microprocessors. This showed up when we traced the supposedly continuous ground for both the power system neutral and the case of the microprocessor, the point where the processor obtained its signal reference. While a shielded isolation transformer was used, the connections were incomplete, leading to an inadvertent ground loop. This defeated the value of the transformer shield as a means of rejecting electrical noise. Remember, we're trying to establish an equipotential signal reference: The same potential for both the power system neutral and the safety (case) ground, with no impedance between the two.

Also, the facility wiring had grown like topsey; thus, we were working with power feeds from several different locations as well as ground wires going in different directions. With the implementation of bonding shield wires and ground locations, an improvement in noise rejection was realized.

MEDICAL SUPPLY COMPANY

We were called into a medical supply company to review its computer installation. The owner wanted a study of prior to the purchase of a UPS. As shown in **Fig. 13.8**, a small 3-phase mainframe computer was connected directly to the utility input power through its own feed and circuit protective devices. The peripherals not only were powered from the mainframe, but also communicated directly with it and with each other. Both the computer and accessories were being operated without any power conditioning equipment.

Paying particular attention to the power circuit from the electric utility, a site analysis was made, with the normal amount of slight irregularities found. Based on this analysis, a UPS was selected to condition and maintain the required power. The installation was completed as shown in **Fig. 13.9**. As can be seen, not only does the UPS supply conditioned power, it acts as a buffer between the mainframe and the utility power, thereby keeping any abnormalities occurring on the line from getting into the computer circuits.

After testing the UPS without load, we provided certification that the new unit was operating in accordance with its performance standards. Since the computer load was very modest relative to the UPS's full load rating, we expected no problems when the computer was initially powered by it. We were in for a surprise!

When the computer's "start" pushbutton was pressed, it began to ramp itself up to operating conditions. Suddenly, the "Power Fail" red light on the mainframe came on. Thinking this was some unusual glitch in the computer's startup system, we began the procedure again—with the same results. This was

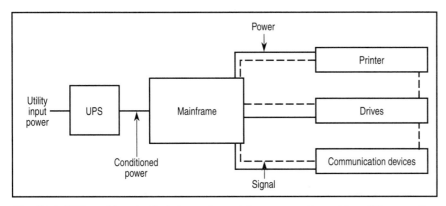

Fig. 13.9. UPS provides a buffer between the mainframe computer and the utility power line.

confirmed with a third try.

Attempting to get to the bottom of the problem, we asked if we might talk to the computer company's field engineer. Specifically, we wanted to know what type of signal system might be sending a message to the "power fail" area of the mainframe. The field engineer was able to identify a certain area of the communications system where electrical noise could interfere with a signal.

This in fact proved to be the case. What was happening was that electrical noise interference in this signal circuit was causing the circuit to tell the mainframe there was something wrong and it should shut down. The source of the problem was signal corruption.

After the field engineer spent 20 min examining the various signal circuits of the mainframe, he found one circuit that was receiving interference from the surroundings. This unprotected circuit was sending instructions back to the mainframe that there was no correct signal interface and, therefore, the computer should shut down. The problem source was identified as a signal wire that was external to the building, run into and out of the earth unprotected from transient noise. With protection installed on this wire, the noise ceased to be transmitted through the signal system. The computer's security check was answered correctly by the signal driver and there was no power fail instruction.

Huge size of this automated warehouse requires the use of lots of steel roof trusses and conveyors, providing a potential for introduction of ground loops into solid-state electronic circuits.

COORDINATION GUIDELINES FOR SENSITIVE ELECTRONIC LOADS

There are many considerations that go into providing suitable power, grounding, communications, and environmental control for a facility that includes significant amounts of sensitive electronic equipment. While each installation has unique requirements, there are many factors that are common to all projects.

The following is a checklist and commentary that can be used to develop a basic approach to providing a power distribution system for sensitive electronic equipment. The list is derived from practical working checklists used by specialists in power quality auditing. More comprehensive working models are available.

CONTINUITY OF PROCESSING OPERATIONS

Before planning the purchase of solid-state electronic data-processing, process-control, communications, workstations, instrumentation, or similar equipment and providing the required power system, it is necessary to first study the operations you plan for the facility and develop some answers to the following questions.

1. What length of downtime can be tolerated in your operation? Many operations can withstand 5 to 20 milliseconds of downtime by their internal ride-through capacity.

2. How frequently can downtime occur? Some locations tolerate one to three power outages in a year's time, but require some additional protection if occurrences total several times a month (20-25 times a year).

3. Are momentary interruptions acceptable? Some hardware and software systems provide for "hold and store" methods of preserving data in case of momentary interruptions.

4. What is the cost in revenue of interruptions and downtime in terms of lost data, "re-boot" time, damaged hardware or software, corrupted intercommunications, and lost processing throughput? The cost of interruption may justify investment in power and grounding improvements.

5. What power-quality related events will interrupt data input, processing, and useful output? For example, even with continuous power to processing hardware, interruption of a separately powered process cooling plant could halt operations. While water cooled processors may be able to stand a few degrees of temperature change, they cannot stand loss of circulation.

6. Are operations dependent upon communication lines that also require continuity of electrical power? The question of whether your telephone system should be on conditioned power must be decided.

7. Is the backup or alternate site operated from the same power source, and are its communications also dependent upon electrical power? You should avoid dependency on common facilities (the same power poles, side of the street, or underground trench) whenever possible.

FACILITY LOCATION CONSIDERATIONS

There are factors over which you have no control that may affect the operation of the planned facility.

1. Will your site be subjected to natural hazards? Storm damage from wind, rain, snow, sleet, ice, lightning, flooding, tidal waves, moving ice, fire, earthquakes, and slides are some of the items that could affect operations. An ice storm can break overhead conductors and topple power distribution poles on a massive scale, and high winds may cause inadvertent breaker operations.

2. Will it be subjected to man-made hazards? Vandalism, sabotage, malicious mischief, arson, collisions from vehicles, riots, insurrections, or war, are all possible causes of interruptions.

3. Will it be subject to offsite damage to support services? Power, communications lines, gas, water, or sewer lines all are vulnerable.

4. Will it be subject to lightning and excavation damage? Even if underground, services damage may occur. If there is more than one power feeder and more than one communications cable to the building, each should be placed in a separate trench rather than in the same trench.

5. Are the local electrical safety and other building codes overly restrictive in the construction materials and assembly methods that are permitted? Unnecessary restrictions can dictate the use of expensive labor, intensive construction materials, and assembly methods. These can increase initial installation costs and continuing maintenance expense in some instances.

COMMERCIAL POWER CONSIDERATIONS

Facilities that will contain high percentages of solid-state electronic equipment are extremely dependent upon a source of dependable electric power. The service record of the utility serving the proposed location should be analyzed.

1. Review power outage and disturbance records of the local utility. Note that some utilities record only "outages," which can be defined as an interruption of power lasting more than 5 minutes. Intermittent breaker operations of 5 to 20 cycles might not normally be recorded. Ask for the number of breaker operations counted on your feeder (per quarter or per year).

2. How does the local utility power quality history compare with those serving other sites? Utilize the site analysis surveys offered by site preparation and coordination firms. These studies indicate what items are typically measured and reported.

3. Ask other users in the area to review with you their power quality records and system operating performance. Be sure to compare each history with the different types of power conditioning devices in use.

4. Examine the support level of the local electrical utility. Many utilities are making available to their customers having lots of equipment based on solid-state electronics, a variety of programs for analysis, design, and equipment selection in order to assist in the enhancement of the power and grounding infrastructure needed to support such equipment. Be sure to investigate and consider this assistance, which may even include equipment leasing or provision of premium quality power supplied directly from a utility-owned power conditioning substation.

5. Be sure to study with the local utility representatives all available options for the normal feeder conditions as well as possible backup feeder availability. Many locations can be served with dual-feeder capability, which can eliminate expensive long-term power support equipment such as UPS, batteries, and engine generators.

6. Can separate distribution lines be fed from separate substations in the event one circuit is accidentally disabled or taken out of service for maintenance? The same strategy should be followed for telephone and data communications cables.

7. Is the cost of electricity excessively high at the proposed location? Compare electricity costs at different locations. Are there demand charge and low power factor penalties? What effect will system efficiency have on actual operating dollars? (A 250kW load would have an energy bill of more than $200,000 per year at $0.10/kWh. Power at $0.075/kWh could save $50,000 per year). Partial load operations will affect long term costs totaling two to three times the apparatus capital expense!

COORDINATION AND PLANNING WITH CPU VENDOR

The characteristics of the solid-state equipment that is to be used varies from one manufacturer to another. It is essential to have data on the specific equipment proposed.

1. Obtain from the selected or prospective equipment vendors their general specifications for overall power quality, grounding, and communications installation.

2. Obtain from the vendors an equipment list containing specifications on:
• line frequency with tolerance limits and maximum rate-of-change;
• line voltage(s), phases, nominal value(s) with upper and lower limits;
• load characteristics for each EDP unit, including kVA, kW, phases, and amperes on each phase for both normal running and starting;
• combined system kVA, kW, power factor, amperes (both rms and peak), and heat output in BTU/hr or equivalent metric units;
• maximum starting inrush currents for units having the ten largest values; and
• site preparation and installation planning manuals.

3. Discuss the proposed system with your electronic equipment vendor to determine the characteristics of not only how well it will operate, but also how it will fail. Concentrate on the general specifications (obtained in 2 above) in determining specific areas of power and grounding enhancement or failure protection, in accord with IEEE P-1100.

4. Determine the conditions to be encountered at the site and obtain agreement from your vendor on the protection that will provide the best operating integrity for the system. Make sure the following items are discussed.

• Power system quality: Providing a constant source of relatively undisturbed electrical power of adequate load capability, suited for the loads, and maintained within limits established for nominal values of line voltage, frequency, waveshape distortion (harmonics), voltage sags, impulse surges, transients, electrical noise, and other attributes of the power source.

• Solid-state electronic equipment susceptibility: Equipment should withstand the disturbances and aberrations of electrical environments in which it is expected to operate, especially the power source.

• Compatibility of load equipment and power sources: This includes harmonizing their respective characteristics, including the interaction between electronic equipment and their power sources. Where source and load both consist of high, dynamic, internal impedances, the risk of distorting the electrical waveshape is high. These distortions can easily lead to power conditioning applications causing more interference than they are supposed to remove.

- Grounding, referencing, and shielding: Providing equipotential signal referencing and shielding that is compatible with grounding requirements for safety.
- Power distribution: Avoiding unwanted coupling by isolation and coordination of power circuits and their respective grounds.
- Conductor sizing and overcurrent protection: Protection from faults and overloads, with special attention to effects of harmonic load currents.
- Lightning and surge suppression/protection: Providing lightning, power switching, and electrostatic discharge (ESD) protection systems.

5. Discuss with the vendor conditions that will make the system fail. The definition of "power interruption" involves time duration. The fact that single-phase AC line voltage varies sinusoidally and passes through zero twice each cycle does not mean there was a power failure every 1/120th second. Electronic equipment is typically designed to store enough energy in its DC filtering elements to ride through each AC zero crossing plus some additional time. Most electronic equipment can withstand a one-half cycle interruption without disturbing the filtered and regulated DC power supplied to logic circuits.

6. Are those responsible for commercial and for electrical wiring and apparatus installation reasonably responsive to the solid-state device vendor's and user's needs? They should be adequately trained and have the necessary technical understanding to work cooperatively with users to help them solve power quality related problems. They must be able to merge noise-control procedures with their construction practices.

DESIGNING THE SYSTEM

Having gathered most of the background information required, the task of actual planning and engineering the power system serving the sensitive equipment can begin.

1. Compare the solid-state electronic unit power requirements with power that is available or can be made available. Remember that voltage transformation can easily be performed with a shielded transformer at the point of use with the added benefit of referring the secondary voltage to a central grounding point, (common-mode voltage isolation). The feeder to the information technology equipment room or area can typically be 480V (or 600V), 3-phase, while the utilization voltages at most of the solid-state devices can typically be 208V or 120V, single phase.

2. If a power peripheral containing an isolating transformer and output circuit breakers is to be used to power other units, the power peripheral will require a branch circuit to supply its input voltage. This can typically be 208, 240, 480, or 600V, 3-phase. The higher voltages are more efficient. They have a lower percentage line voltage drop and generally cost less to install for a given kVA rating. In this type of installation, the secondary output circuit breakers and conductors to unit loads are not called branch circuits because they are system interconnections rather than part of the building wiring. As part of the sensitive system, the power peripherals and their connecting cables are subject to UL 478 examination and listing rather than inspection under NFPA 70 or other applicable electrical codes for building wiring.

3. Itemize the loads and draft a wiring connection schedule.
- Place each sensitive load that draws more than 5A on a separate power peripheral circuit with its own circuit breaker and interconnecting power cable.
- Arrange single-phase loads so they will be evenly distributed over the three phases and neutral.
- All electronic loads need not use the same power source to avoid ground potential differences. Using separate shielded isolating transformers, some loads may be operated from UPS output, while other less critical peripherals (printers, for example) could operate on commercial power.
- Separate shielded isolating transformers may be used to separate easily disturbed loads (memory, for example).
- Be aware of any loads that have a DC load component of current such as created by a load with half-wave or unsymmetrical rectification or SCR control.
- Be aware, also, of loads having a very high harmonic content in their load currents. These can create operating problems due to saturation or overheating the first upstream transformer or MG set. Besides high temperature rise, the observable effects may be: excessive operating costs, poor efficiency, and distorted output voltage and input current waves. Such effects can render UPSs inoperative and can cause very high peak magnetization current pulses at the inputs to some distribution transformers. Refer such problems to the manufacturer first to see what can be done to eliminate or reduce the problem. Otherwise special line conditioning equipment may be required. Single-phase harmonics may be additive in the neutral circuit by as much as 173% of phase currents

4. Examine site layout proposals for environmental compatibility.
- Have process cooling locations been placed and sized with respect to heat producing areas on the floor?
- Have present and future underfloor space needs been addressed?
- What flexibility has been planned for expansion of the power, grounding, or communications protection systems?
- Has a central operator station been planned for environmental control or access?
- Has the routing, type, and physical installation of data communications been planned for both present and future compatibility with the site?

5. Since some equipment vendors may have included power conditioning internally in their processing units, you need to incorporate these considerations in your overall power quality implementation. Many power conditioning techniques will interact one with the other to create "flip-flop" distortion. Make extensive examination of each device intended for the system to determine internal impedance. Proper coordination will be achieved when equipment with high dynamic impedance is interfaced only with the equipment with low dynamic impedance. In

this way, the distortions occurring between highly solid-state power supplies are softened and disruptive interaction can be reduced.

6. Incorporate unit cabling restrictions with room layout plans. Organize each unit so it will not interact physically with surrounding sensitive units (such as printer dust with open disk drives).

7. Review data communication installation requirements to insure long term reliability and protection to minimize data errors. Be sure there is no signal paths susceptible to noise contamination that might produce error signals to "powerfail" the processor.

8. Be sure to direct your designers in the planning stages, to have HVAC equipment supplied by a transformer and feeders other than the one(s) used for the solid-state electronic equipment. Processing power should be provided by separate transformers and feeders, not shared with other loads.

MATCHING SYSTEM POWER REQUIREMENTS TO POWER CONDITIONING ALTERNATIVES

Part of the design process involves determining the answer to questions affecting the power requirements.

1. What capacity is needed in kVA, kW?

• The sum total of power in a specific equipment list is the "connected load." However, actual measured power use may be less because each unit may not have all options installed and may not draw maximum connected load continuously or simultaneously.

• Make allowances for future system growth. Power requirements for memory and the controllers, multiplexers, and exchanges needed to address and share large memory may grow faster than the rest of the system.

• Divide total power in watts by the floor space devoted to processing use. Typical large systems use 50 to 60W/ft2. A 50´100 foot room might use 250kVA of power, for example. If the proposed installation should be substantially more or less than this energy density, it may be worthwhile to verify the reasons.

2. Are the internal impedances and momentary overload characteristics of the power source adequate to handle the short term demands without installing more capacity than needed for the steady-state load? The internal impedances should be low enough so that transient currents will not create excessive load-induced line voltage disturbances.

3. Has verification been made of the operating efficiency under partial load operation (partial load efficiency)?

The cost of additional electric energy for a drop of 15% in efficiency (say from 88% to 73%) can be in excess of $10,000 per year, for every 100kW of load!

The estimated total load should be compared with the expected steady-state actual running load of the system and the result used in determining the "partial load" ratio as a percentage of the nameplate kVA/kW of the power conditioning device. Rarely are systems installed that run higher than 50 to 60% of the nameplate capacity of the power source!

In order to assure optimum efficiency, be sure to request manufacturer's guarantees (in writing) at 50% load operations, with the guarantee to compensate users for operations not meeting those standards.

4. What "ride-through" and other power conditioning continuity requirements, if any, are necessary to satisfy the operating conditions of the solid-state electronic equipment?

• Without supplementary energy storage devices, many units will ride through a 5 to 20 millisecond interruption of power without malfunction, provided noise impulses associated with the interruption do not reach and corrupt digital signals by other paths.

• Ferroresonant and synthesizer transformers may enhance ride-through slightly in some cases, provided the loads are not sensitive to phase shifts during their correction of line voltage variation, and loads can be limited to 75% of device rating.

• Motor-generators can extend ride-through to as long as 20 seconds. Check the equipment manufacturer to determine whether synchronous 60 Hz power is needed or whether induction motor drives with lower (varying) frequency of output can be tolerated.

• UPSs can extend ride-through to typically 5 to 30 minutes or longer.

• UPS installation with emergency standby diesel power can extend ride-through almost indefinitely.

• Examine the power conditioner or independent power source for its limits on load current, kVA, and kW output. This will vary with the power factor of the load. Some devices such as static UPSs have very little overload capability and depend upon a stiff, low impedance bypass power source (usually the public utility) to supply large starting loads. Verify that the power source and any bypass will supply all inrush needs and still provide the power quality that is needed.

• If voltage regulators are used, verify that their response times are fast enough to follow line voltage changes, yet will not interact with regulators in the sensitive loads.

• Determine if a step-by-step approach to power conditioning will benefit your installation. For optimum lowest dollar outlay, consider first a power conditioner with provisions for in-the-field conversion to battery back-up support at a later date, without loss of initial investment. Recent developments in technologies provide installations at 50% of UPS dollars that perform at 95 to 98% of UPS levels.

5. What are the predicted line voltage sags, swells, and impulse transients? It is important to note that ordinary switching of loads can be expected to create momentary impulse voltages as high as the peak value of the sine wave.

6. What are the significant sources of load-induced transients? Consider the following ideas to minimizing their effects:specify and order "soft start" system operations; put units with high inrush on separate circuits; and put sensitive units and other units that create disturbances on separate shielded isolating transformers.

GROUNDING FOR CONSISTENT NOISE SUPPRESSION

One of the most important item to be considered in any design is the safety and signal grounding required.

1. Are grounding requirements of the solid-state electronics manufacturer

consistent with FIPS Pub 94 and/or the IEEE Emerald book? Differences and underlying rationale should be discussed and agreed to. There may be good reasons for differences.

2. Is a shielded transformer required or recommended? This affects the point where logic ground conductors and power source neutral grounding points will come together at a common point.

3. Will all conductors (power, communications, and grounding) be brought to the point of delivery to your system through one very close-coupled "entrance" or "window?" Scattering the entrances and exits for the wiring increases the risk of noise voltages and transient impulses circulating through the system.

4. Where will the system's central grounding point be located? If a modular power center (power peripheral) is used, it may be located there.

5. Will the information technology equipment room raised floor structure be specified with interconnecting bolted horizontal struts, suitable for use as a signal reference grid? This could save much money compared with construction using copper conductors or straps, and could enhance performance.

6. Are the ground conductors for nonsensitive equipment separated from solid-state electronic equipment grounding conductors except at some upstream common connection, typically at the building service equipment or other common separately derived power source (such as a transformer)?

7. Will the communications and power grounding systems in the building be bonded together at an appropriate upstream common point? This is needed for safety and to minimize noise voltage differences without providing conducting paths through the ground conductors of the sensitive systems.

8. Are all grounding conductors and conducting pipes that penetrate the sensitive area bonded together by short, robust connections before they enter the information technology equipment room?

9. Are all sensitive units and their accessories listed or approved by UL or other acceptable safety testing laboratory that is recognized by the municipality in which the units are to be installed?

10. Does the premises wiring meet the local/national electrical code requirements?

11. Does the installation comply with the UL listing by the manufacturer?

REDUNDANCY REQUIREMENTS

Having designed a basic power system for the sensitive equipment, the need to further protect the system by building in redundancy must be explored.

1. Perform a simple failure analysis by assuming that each piece of equipment, its power source, wiring, and wiring devices can fail or must be deenergized for service.

• Determine if the electronic systems will continue to function and recover in such instances.

• Determine if redundant bypass paths in the power sources around conditioning equipment and major pieces of electrical apparatus will provide the ability to continue in event of failure, service, or replacement.

2. Review the above in light of probable failure rates and frequency of necessary maintenance.

3. Should margin for future growth be supplied by redundant smaller units rather than initially selecting a single unit adequate for all future growth?

4. Is there sufficient floor space and HVAC capacity to enable a new system to be installed and to become operational before dismantling the existing unit? Reserve amounts of power may not require power conditioning for this purpose, but once a new system is operational, there is the problem of transferring the power source with a minimum interruption of service.

DATA COMMUNICATIONS PROTECTION

The ability to transfer data and signals from one point to another within the facility or transmitting to another location is essential.

1. What level of security is to be provided for data lines entering an ITE room or area?

2. Has consideration been given to routing of data lines with regard to nearby induced interference?

3. Will power and communication grounding be done to reject noise from the data lines?

4. What level of protection has been incorporated in data lines leaving the building?

5. What techniques for shield/protection have been approved by the solid-state device manufacturer?

6. Is the installer qualified in the installation and termination of the specific cable system selected?

LIGHTNING PROTECTION

Protection of the installation from transient voltages introduced by lightning can be a major factor in areas that have a high incidence of thunderstorms.

1. Does the building structure have lightning protection? Refer to UL Subject 96A and Lightning Code NFPA 78.

2. A building in which structural steel is bonded together by welding (as opposed to reinforcing steel in concrete, which is electrically discontinuous or merely touching) offers better lightning protection of circuits within the building.

3. Conductive parts from roof-mounted equipment (lightning rods, etc.) should not create a path to ground via sensitive circuits. Air conditioning coolant pipes from the roof to air handlers must not become a direct path for lightning to reach critical circuits, or central grounding points. All metallic runs should enter and exit at a single access point where all parts are bonded together and grounded to building steel.

4. Are lightning protection conductors to ground separated by at lease 6 feet from power of communications circuits to avoid induced noise? They must be taken to ground by a separate path from the equipment grounding system.

5. All incoming power and communications equipment conductors should be protected by surge protection devices having shunt overvoltage paths to ground, and series impedances to limit surge currents. The place for secondary lightning protection is at the building entrance. Supplementary protection should be placed at the input and output of load devices such as rectifier/chargers for UPS installations, motor-generators, and voltage regulators, as well as at the load devices themselves, and their communication ports.

6. Does secondary lightning protection include wave front modification? The installation should use the most efficient "3-stage" transient suppression. The first part of secondary protection is necessary at the building service entrance. Here the large block of energy must be dissipated to the building service ground, without interference with solid-state electronics. The second part of this protection, at downstream locations is to protect from low-level voltage transients that may pass through the large energy protector. In order to achieve complete protection, it is critical that these two separate applications be planned together to achieve a coordinated approach capable of handing all lightning surge levels.